"This rare and potent blending of voices highli[...] the Dalai Lama as they share deep wisdom an[...] times. Their plea goes right to the heart: Out o[...] on behalf of our precious planet. It's not too late!"

—Tara Brach, author of *Radical Acceptance* and *Radical Compassion*

"The crisis with our planet can feel overwhelming, which is why I am so grateful for this remarkable book. Grounded in the conversation between the Dalai Lama and Greta Thunberg, with other valuable reflections added from various experts, the result is a resource that is deep, practical, and timely. Read this book, take notes, and then go and make a difference."

—The Most Rev. Michael B. Curry, Presiding Bishop of The Episcopal Church and author of *Love Is the Way*

"We are in a moment when very ancient wisdom and the latest science are converging on some difficult and important questions for our species—and this conversation between young and old is a perfect example of how rich this moment can be!"

—Bill McKibben, author of *The End of Nature*

"In *A Future We Can Love* Susan Bauer-Wu has created a captivating chorus of voices from key environmental thinkers articulating the dire realities and hopeful actions we can each take to stave off the climate crisis. Let's hope a future we can love becomes the future we might have."

—Daniel Goleman, coauthor of *Why We Meditate* and *Altered Traits*

"Do we have the will, the capacity, and know the action to take to protect our earth from climate change? Inspired by a conversation between the Dalai Lama and Greta Thunberg, *A Future We Can Love* is a profound call to action. Rather than be an abstract or philosophical declaration, it provides tangible ideas to help support our Earth for future generations."

—Sharon Salzberg, author of *Lovingkindness* and *Real Happiness*

"So many of us want to look away when it comes to this massive global problem. But this book pulls off the neat and crucial trick of bringing all of us into the conversation and giving us the tools to help out, without falling into the traps of resignation, avoidance, or denial."

—Dan Harris, author of *10% Happier* and host of the
Ten Percent Happier podcast

"Enter and be entrained into an inspiring and illuminating conversation and inquiry—one that our very lives, all of us and all life on Earth, depends on. May it lead to potent transformations inwardly and outwardly at all levels of scale."

—Jon Kabat-Zinn, founder of MBSR;
author of *Coming to Our Senses* and *Full Catastrophe Living*

"*A Future We Can Love* is a profoundly moving personal reflection on the climate crisis. The book provides an unflinching overview of current science, then embraces a roadmap to solutions. How we shop, eat, and vote—our climate values matter. Susan Bauer-Wu interviews world-renowned climate scientists and augments their sobering facts with a healing vision. She shows us how to navigate these challenging times with compassion and love. In a world of eco-anxiety, Bauer-Wu's wisdom inspires respect for our planet and all living things, along with a much-needed sense of hope and possibility."

—Roberta Baskin, investigative journalist and climate activist

"*A Future We Can Love* is a rare weave of factual clarity, passionate outrage, deep teachings, charming stories, personal practices, and lovely, lovely heartfelt prose. The book is a gem—the best I've seen both introducing people to the bad news while also emphasizing what we can actually do to create the future we love."

—Rick Hanson, PhD, author of *Making Great Relationships*

"Never before have I seen the truths of our climate situation presented in such an intimate and vulnerable way. Susan's conversation allows the world's greatest scientists and spiritual leaders to become our personal guides. Together they give expression to our heartbreak, and together they describe the path back to hope . . . and to action. A must-read!"

—Philip Clayton, author of *The New Possible*;
president, EcoCiv.org

A FUTURE WE CAN LOVE

EFFECTIVE APPROACHES TO THE CLIMATE CRISIS THAT BEGIN WITH US

FEATURING CONVERSATIONS WITH LEADING CLIMATE SCIENTISTS, HIS HOLINESS THE DALAI LAMA, GRETA THUNBERG, AND OTHER SPIRITUAL TEACHERS AND CHANGEMAKERS

SUSAN BAUER-WU

with Stephanie Higgs

SHAMBHALA

Shambhala Publications, Inc.
2129 13th Street
Boulder, Colorado 80302
www.shambhala.com

Except where otherwise noted, quotations in this book come from talks
delivered at Mind & Life conferences in 2021 and 2022, the Climate Emergency:
Feedback Loops films (feedbackloopsclimate.com), or from conversations with
the author. Occasionally these have been edited for length and clarity.

All author proceeds from this book go to the Mind & Life Institute,
a nonprofit organization based in the United States.

Cover art: Yuriy2012/Shutterstock and VVadi4ka/Shutterstock
Cover design: Daniel Urban-Brown
Interior design: Lora Zorian

9 8 7 6 5 4 3 2 1

First Paperback Edition
Printed in the United States of America

This book is printed on acid-free, recycled paper.
Shambhala Publications is distributed worldwide by
Penguin Random House, Inc., and its subsidiaries.

LIBRARY OF CONGRESS CATALOGING-IN-PUBLICATION DATA
Names: Bauer-Wu, Susan, author. | Higgs, Stephanie, author.
Title: A future we can love: effective approaches to the climate crisis
that begin with us: featuring conversations with leading climate
scientists, His Holiness the Dalai Lama, Greta Thunberg, and other
spiritual teachers and changemakers / Susan Bauer-Wu with Stephanie Higgs.
Description: First paperback edition. | Boulder, Colorado: Shambhala
Publications, [2024] | Includes bibliographical references and index.
Identifiers: LCCN 2024003616 | ISBN 9781645473527 (trade paperback)
Subjects: LCSH: Environmental protection—Religious aspects—Buddhism. |
Climatic changes—Religious aspects—Buddhism.
Classification: LCC BQ4570.E58 B38 2024 | DDC 294.3/927—dc23/eng/20240509
LC record available at https://lccn.loc.gov/2024003616

With boundless gratitude to His Holiness and Greta

CONTENTS

Now, She still ripples.
She still hums, pulses, quivers.
She still sighs,
murmurs under the Skies.

WE PAY ATTENTION, and all we hear is urgency. Waters whirl, winds rise, fires rage, irate. The challenges are innumerable, but also infinite are the opportunities. Our grief is daunting, but also heartening is our compassion. We course a cosmic webbing holding awe and horror, wonder and doubt, creation and transition . . . us and all others.

This abysmal relentless weaving is love in all its myriad forms.

We listen whole our Mother Earth's humming, Her calling, Her heartbeat throbbing, and Her ails. We, as made of soil ourselves, are porous. Enacting love flows—throughout— quenching the cracked soils of hopelessness, helplessness, and isolation.

Breathing in, we return, expanding full gratitude.

Breathing out, we connect, unfolding kindness and care.

It is love who guides grief to meaning, anger to action, despair to transformation, fear to safety. Thus, from love, all injuries heal; they repair, restore . . . and bridges *open.*

Because our beings—whole—*open.*

Trust becomes.

Relatives, recall those early steps of unknowing and discovering!

Yes.

Those first steps we walk again right here, right now.

Today, we walk our steps attentive and intentional. Our past brings forth sensible alertness now. Tomorrow is right here— made of us—right now.

Yes.

Bring to heart the time we walked barefoot. When our feet caressed the skin of our Lands, concerned little of thorns and pebbles, seeking first to play and connect.

Relatives, evoke the gentle holding of our Mother Earth, Her caring gaze, and Her smiling.

We smile back because we are indeed listening.

> *Now, we ripple.*
> *We hum, pulse, quiver.*
> *We sigh,*
> *murmur under the Skies.*
>
> —YURIA CELIDWEN

A FUTURE WE CAN LOVE

INTRODUCTION

A BUDDHIST LAMA AND A CLIMATE ACTIVIST walk into a Zoom . . . It could be the beginning of a joke, but this is not a joke. The lama was the Dalai Lama, the activist was Greta Thunberg, and the conversation really happened. The reason for their meeting—the climate crisis we are living in—could hardly be more serious. Like anything having to do with the Dalai Lama, though, the conversation wasn't joyless. There was energy and levity, even given the heaviness of the subject matter and the extremely early hour for Greta—the middle of the night, really—in Sweden. It was late where I work, too; in Charlottesville, Virginia, it was 10:30 at night, but I'm a night owl, and anyway, it could have been any time of day or night and I would have been just as wide awake and pleased to be there, even given the heaviness of the subject matter.

January 10, 2021, nine o'clock in the morning in Dharamshala, India, and the Dalai Lama and Greta Thunberg were meeting for the first time. Nearly a million people had tuned in for the livestream to hear what these two global leaders had to say to each other about the climate crisis. I was there to introduce the event, since the organization I work for, the Mind & Life Institute, was hosting it. I had met Greta about an hour before (3:45 a.m. her time, I'd like to acknowledge), for sound check. At first, she was maybe, understandably, a bit sleepy, but someone, I think her father, brought her tea and toast, and she quickly turned into the clear-eyed and preternaturally composed young person I recognized from the internet. Dressed in a black hoodie with her hair pulled back, she was in the living room of her family home, the background refreshingly unstaged. A rumpled throw was on the sofa, and a leftover Santa hat was perched on—was that a hat stand? a lamp?—though it was weeks since Christmas. Greta showed up with no pretenses, just her trademark sincerity about the state of the planet. I could see the Swedish midwinter pitch dark outside the window behind her.

The Dalai Lama smiled and waved to us from a comfortable chair behind a small wooden table with a clock on it. He appeared in a sunlit room full of maroon and yellow flowers, which picked up the maroon and yellow of his traditional monastic robes. He has been advocating for the environment for many decades, since long before Greta was born, back when the hole in the ozone was more widely known than the threat of humans' effect on climate. But just the previous year, the Dalai Lama had sent Greta a letter. "I am also an ardent supporter of environmental protection," he wrote to her. "We humans are the only species with the power to destroy the earth as we know it. Yet, if we have the capacity to destroy the earth, so too do we have the capacity to protect it. It is encouraging to see how you have opened the eyes of the world to the urgency to protect our planet, our only home. At the same time, you have inspired so many young brothers and sisters to join in this movement."

Now, to Greta, the Dalai Lama reiterates the admiration and optimism that had inspired him to write that letter, and he says he is eager to hear what she has to say. "Younger members of humanity are showing a genuine sense of concern for our future, for our planet, and this is a very, very hopeful sign," he tells her. In her reply, her first words to the Dalai Lama in person, the appreciation is mutual: "I can say as a younger generation we are eternally grateful that you are standing up for us, not only for us but for the future of the entire humanity and for the entire planet."

Greta often points out, and rightly so, that it's neither fair nor helpful to put all our hope in young people and leave it to them to save the planet; by the time she and her peers are old enough to be environmental scientists, climate journalists, elected officials, or green engineers, we will have missed the critical window in which to avoid catastrophes associated with global warming above 1.5 or 2 degrees.[1] Naturally she appreciates that, unlike most grown-ups, the Dalai Lama has been active on behalf of the planet for much of his long life. He sleeps a solid nine hours a night, but when it comes to the environment, as the saying goes, he didn't get out of bed just yesterday.

His Holiness is also right; the youth-led movements that captured the world's attention *are* encouraging, and we should take hope where we can find it. It occurs to me, as I listen to them, that somewhere between the eighty-five-year-old Tibetan Buddhist leader and the eighteen-year-old activist, between the sage legacy and the whole life ahead, is where the rest of us must come in.

And we must. Most of us know this at some level, however recently the knowledge has come to us or started to sink in, whether received in news headlines or documentaries or more concrete and immediate forms like heat waves, wildfires, floods, or water shortages. Most of us have been told the bad news about climate change, and most of us believe it, whether or not we've allowed ourselves to think or do much about it. These days we hear it again and again. Many people around the world are living it. But what's different about these two people, Greta and the Dalai Lama, coming together to talk about the climate crisis is that they make the conversation feel welcoming. And this could change the game. At least it did for me.

WE MAY NOT WANT TO TALK about it, but lately we would have to go out of our way to avoid seeing that our planet is facing something more dangerous than we humans have ever collectively seen, recorded, or experienced. While extreme weather events aren't new, the accelerating pattern and severity of them are alarming. The news of weather-associated disasters is coming at me so fast that I have begun to expect them somewhere in the world each week, if not each day. The constant drip, drip, drip of terrible headlines can begin to feel normal, or at least numbing. I watch the news, but I don't really "see" it. I hear about a mudslide or a fire that destroys a town and I shake my head, yet the suffering feels too great to bear, so I turn away. It isn't that I don't care but that I feel helpless to do anything. Or I did, until the people you will meet in this book helped me see that I am actually connected to all of it, that there are already people doing something about it and I can join them.

Perhaps the climate crisis feels like something that happens to "other people," far away. Maybe it seems it is something that will happen in the future, giving us time to change. Or perhaps we fear that if we allow ourselves to feel the suffering of the world, we will despair or be paralyzed. How would we get out of bed in the morning if we actually felt the loss and pain of what we're doing to one another, to countless nonhuman beings, and to the planet, even though we now know better? Perhaps we think or pray someone else will take care of it, step in at the last minute with a technological miracle or ride in on a white horse in such charismatic style that suddenly everyone gets on board. Charismatic leaders and technological innovations can help, but they alone can't save us.

Many of us do what we can and what we're encouraged to do individually—recycling, supporting pro-environment candidates, installing solar panels or buying an electric car if we can afford it. Then we're confused and disheartened when we hear, often from well-meaning people, that our individual choices don't matter. We carry on recycling and eschewing plastic straws, but the truth of our climate emergency still haunts us. We feel puny compared to rising oceans and melting polar ice caps. But it's simply not the case that the inadequacy of individual effort means that our best option is quiescent resignation. Uncomfortable though it may be, we need to learn to talk to one another about the climate crisis.

Greta said to the Dalai Lama, "It was and is my experience that there is a huge lack of awareness of the true issue of climate change and the risks that arise from this, and that we as a society spend too little time discussing it. And the discussions that take place are too narrowly focused. This primarily occurs because science is not involved enough." Research concurs that while most of us are alarmed or concerned about the climate crisis, we're not talking about it. Maybe that's because we think we don't know enough and we're afraid to learn, afraid of what we'll find out; or because we know enough to be too afraid to say it out loud and we'd rather push it down; or because it can be exquisitely awkward. The more we know, the more it seems like a socially unacceptably dark place to go. I know I'm not alone in not wanting to "go there."

For a long time after I began hearing about climate change, I carried on with my life and just tried to be a decent person, doing my bit of good in the world and clinging to normal. Then a Mind & Life colleague and filmmaker, Barry Hershey, was struck by the power of climate feedback loops—processes by which heating the planet can lead to more heating—and decided to make a series of short films about them. As he told me more about the films, I was taken and wanted Mind & Life to help launch them. We began to make a plan for a launch event in January 2021 with the vision to include the Dalai Lama and Greta. His Holiness cofounded the Mind & Life Institute thirty-five years ago and continues to stay engaged in our organization's work. He was keen to have a live public conversation with Greta. She was a longer shot, but given the quality of the films, the opportunity to meet the Dalai Lama, and the potential for the meeting to affect millions, we decided to reach out.

But getting in touch with her was a challenge. After unsuccessful attempts to contact her through intermediaries, we learned that the Dalai Lama had written her that letter and hoped that reminding her would encourage her to say yes. Everyone involved in planning the program was on tenterhooks as the clock ticked through the end of November and into December. Then, less than a month before the event, Greta confirmed. We all breathed a big sigh of hooray! We worked through the holidays to iron out the details for the event.

WHEN THE DALAI LAMA PARTICIPATED in the first Mind & Life Dialogue, in 1987, he was already some fifty years into his remarkable life. Born to a farming family in the traditional Tibetan region of Amdo in 1935, he recalls the Tibet he grew up in as a "wildlife paradise" ("No exaggeration").[2] He can remember traveling from Takster, his birthplace in eastern Tibet, to the capital city of Lhasa where, as a four-year-old boy, he was formally proclaimed the Dalai Lama. In particular, he remembers the wildlife that captivated him along the way: "Immense herds of kiang (wild asses) and drong (wild yak) freely roamed the great plains. Occasionally we would catch sight of shimmering herds of gowa, the shy Tibetan gazelle, of shawachukar, the white-lipped deer, or of tsoe, our majestic antelope. I remember, too, my fascination for the little chibi, or pika, which would congregate on grassy areas. They were so friendly. I loved to watch the birds: the dignified gho (the bearded eagle) soaring high above monasteries and perched up in the mountains; the flocks of geese (nangbar); and occasionally, at night, to hear the call of the wookpa (the long-eared owl)." He remembers that even in Lhasa he did not feel "in any way cut off from the natural world."[3]

Twenty years after his idyllic childhood journey, in 1959, the Dalai Lama was forced out of his homeland. Tibetans who have been back to visit since then report to him the environmental devastation they have seen—the conspicuous loss of wildlife, the razing of forests—"clean-shaven like a monk's head" is the Dalai Lama's impression from talking to them.[4] He has long been aware of the connection between this deforestation at the source of many of Asia's great rivers and flooding in places such as Bangladesh. Ten or fifteen years ago, a Chinese ecologist introduced him to the idea of the Tibetan Plateau as the "Third Pole,"[5] an expanse of ice so large and consequential and, like the Arctic and Antarctic, warming so much faster than the rest of the globe that we best pay attention and do a better job of protecting it.

For more than sixty years, the Dalai Lama has led his people in exile—refugees forced to leave behind everything they knew—as waves of climate refugees are having to do now. So many people have already been pushed off desertified lands, with two hundred million climate refugees anticipated by 2050 (some estimates go as high as a billion).[6] The story His Holiness tells, and has lived with courage and grace, has much to teach us about what can sustain us through that kind of loss; about how to preserve our humanity even through periods of darkness, trauma, despair. The Dalai Lama was awarded the Nobel Peace Prize in 1989. He was the first recipient whose citation emphasized work to conserve the environment.

"When I was in Tibet, until I was twenty-four" he said in talks at an environmental summit in 2013, "whenever we passed through rivers or streams, we always enjoyed it. Only after I came to India did I first hear 'you cannot drink this water,' and I was really surprised. Then I began to learn that even though water looks like water, often it is still much polluted. Fish and other water animals sometimes have difficulty to live. But I remember that one time in Stockholm, in the center of the city there is one river, some of my friends mentioned that now fish in the river began to reappear because of the special care about pollution. Although factories are still there, they had made some special effort to not pollute the water. As a result, some fish had gradually appeared. Previously, there were no fish. Therefore, gradually I developed a keen interest about environmental issues."[7] Stockholm! When I see him face to face with Greta, I enjoy that coincidence.

Up in the middle of the night. A sacrifice, as so much of her work has been. Seven days before she met the Dalai Lama, Greta celebrated her eighteenth birthday. She would be returning to school the following Monday after a long time away. In public interviews and statements, she often makes it clear that she would much prefer to be in school, that she is sacrificing her education and her childhood to wake up the adults. Not for them to "tell us what you consider is politically possible in the society you have created," as she said to the British Parliament, "but for you to put your differences aside and start acting as you would in a crisis. We children are doing this because we want our hopes and dreams back."[8] We apologized for intruding on her sleep.

In our planning for what I've come to call "the Conversation," I thought a lot about the contrast between the backgrounds of these two leaders of the climate movement. Where the Dalai Lama was born into a virtually premodern Tibet and has traversed centuries of technological development in his lifetime, Greta is a child of urban, industrial, postmodern Sweden. Born in 2003—the same year camera phones became widespread—she is a Gen Zer who never saw a moment of the twentieth century. Like so many her age, she is fluent in social media (with five million Twitter followers as I write this) and uses it to keep a global climate conversation going. But even in Greta's relatively short time on Earth so far, she has seen climate change affecting her homeland—change that is parallel to what is happening to the Tibetan Plateau. One-sixth of Sweden is north of the Arctic Circle. Greta has traveled north herself, with her father in their electric car, to see how the tree line has crept up the mountains and how the Alpine zone is shrinking, forcing the animals that live there higher and higher "until they have nowhere else to go";[9] and to hear from the scientists in the field there measuring these changes how the rate of change is accelerating.

The Conversation turned into quite a project, and as president of Mind & Life, I became responsible for that project. In the process, the feedback loops and the climate crisis became something I couldn't ignore, and what's more I didn't want to.

There was a lot to accomplish—as is so often the case with the projects the Dalai Lama is passionate about; and there is much to say about the Conversation. There were climate scientists there, too, to answer questions from these two famously, fearlessly curious people, and to explain the science to all of us listening. I can't wait to introduce you. In these pages you'll meet the permafrost expert Susan (Sue) Natali, the five-time Intergovernmental Panel on Climate Change (IPCC) report lead author William (Bill) Moomaw, and many of their colleagues (who also participated in Barry Hershey's films). As Greta said in the event, affirming Barry's documentarian instincts, "We cannot solve the climate emergency without taking these feedback loops into account and without really understanding them. So that is a crucial step." The films are online, with subtitles in thirty-two languages, to watch for free (www.feedbackloopsclimate.com).

But beyond the opportunity to learn more about climate feedback loops, something else happened that day (or night, depending on your time zone). The very fact that His Holiness the Dalai Lama and Greta Thunberg were in a room together, talking about the greatest challenge humans have ever faced, brought light to the darkness. Feeling some of my own fear melt away, I realized that for the first time, I *wanted* to go there. I wanted to be a part of this conversation. And I wanted to invite others to join us.

This book is my invitation to you to join Greta and the Dalai Lama and a whole bunch of other thoughtful and concerned people to talk about a future we can love instead of fear. As Greta said to the Dalai Lama, "Although we may be very different in terms of age span and many other things, we share the same goal; we share a common goal and that is to protect our planet and life on Earth and humanity." Who doesn't share this goal? Let's talk about what it would take to realize it.

THERE HAVE ALWAYS BEEN PEOPLE in our world who are willing to take on seemingly impossible odds, risking or sacrificing their own well-being for the benefit of others. They see suffering, injustice, or danger and run toward it in order to sound an alarm or pull others from the jaws of harm. They do not shrink in the face of structural barriers, challenges, or setbacks, let alone social awkwardness. These are our heroes, our saints, our leaders who have the integrity and courage to take on the status quo and call out those who would do us harm. In the Buddhist tradition, such people are called *bodhisattvas*, people who see the true nature of what is and are willing to do whatever it takes to relieve every single being of their suffering. It is said they never give up on anyone. When we hear their stories or see them at work, their actions thrill our hearts. The Buddhist tradition also holds that we are all innately bodhisattvas, that we each are born with this capacity for compassionate action. The question isn't, What would the Dalai Lama or Greta Thunberg or Sue Natali or Bill Moomaw do. The questions are: What will I do? What will you do? What will we do?

Greta, the Dalai Lama, and the scientists represent everything necessary to address the issues before us. As Greta has said, "We already have all the facts and the solutions" we need.[10] As the Dalai Lama said that day, "Our way of thinking needs to re-orient itself."—which is what he has been saying his whole long life about changes that start from seeing what *is* rather than what we would prefer to see. This is why even people who only heard about their conversation felt a little better about the world just knowing it happened.

Thanks to the enormous amount of work scientists have done, today we can no longer claim ignorance about the crisis we're facing. Fortunately, science is also showing us, with increasing specificity, what collective action we humans need to take if we are to meet this challenge. The Dalai Lama is calling on our collective human conscience to embrace our common humanity and our compassion for future humanity, based on deeper appreciation of the interconnectedness of the welfare of all beings that share this fragile blue planet. Greta and all the voices that have joined hers are saying, with the science behind them, Okay, so *let's do this*. Let's do everything we can now for a future we can love.

THE DALAI LAMA SAYS if he hadn't become a monk, he would have been a scientist or an engineer. The Mind & Life Institute exists, with his ongoing support, to foster conversation between scientists, spiritual teachers, and changemakers, based on the idea that they can learn from one another, seek new directions together, and generally bring out the best in one another for the sake of positive change in the world. We've been doing this for thirty-five years, as I said, but since I joined the organization as president in 2015, we've expanded our mission from promoting "human flourishing" to "flourishing"—of all life, all beings—in recognition of the crucial truth that we flourish *with* nature—or we don't. Whichever way it goes, we go together. We're part of nature, inseparable, and too many people seem to have forgotten this interdependence.

Although it's the Mind & Life Institute's unique ability to bring together diverse voices on research, contemplative wisdom, and action that makes possible the rich conversation you'll find in this book, I'm writing to you now not so much as president of that organization but as me, Susan Bauer-Wu—human, mother, grandmother, and worried citizen of this planet. Since that winter day in early 2021 when the Dalai Lama and Greta met, I've put solar panels on my house, participated in a "council on the uncertain human future,"[11] and taken many, many fewer flights than usual, not only because of the COVID-19 pandemic. With my newfound not fearlessness exactly but climate honesty, solidarity, and courage, I've been catching up on my reading about the climate crisis, listening to podcasts on the subject, and talking with everyone I know about it; and I am fortunate in my work and life to know many, many knowledgeable, wise, and compassionate people. This book came out of all these conversations, and I thought it would be fitting if it felt like a conversation—not me telling you about climate change, because I'm not qualified to do that. Just us talking. Because I want more people to know what that can look and feel like.

I am qualified and blessed to be able to bring people together in conversation. The people you will meet in these pages weren't all in a room together at the same time. A few of them aren't even alive. But I've taken the liberty to write it as if we were all together, in the present tense, because that's been my experience since that time with Greta and the Dalai Lama. And a growing, expanding conversation is what we need and what our planet needs from us at this time.

Everyday conversations don't typically begin with a thesis or a preset agenda, but a book needs structure to hang on and an argument for a reason to exist. The argument that emerged from this conversation—especially with Thupten Jinpa, my friend, Mind & Life colleague, religious studies scholar, and the Dalai Lama's longtime English translator—is this: the path to a future we can love starts with knowledge and proceeds, through our capacity and willingness to change, to action. Hence the four-part structure of this book:

- knowledge
- capacity
- will
- action

This sequence can be derived from Buddhist theory of ethics and karma, and from the basic assumptions of pretty much every system of ethics or law.

Responsibility starts with knowledge. When, in her conversation with the Dalai Lama, Greta was invited to give a call to action, she said to everyone listening, "If I could ask one thing of you, it would be to educate yourself, to try to learn as much as you possibly can." The Dalai Lama agreed, adding, "Much depends on education." We need enough knowledge to act responsibly and to be responsible for our actions. And perhaps the most crucial form of knowledge is to know our own capacity, individually and collectively, as well as the planet's capacity to heal. We *can* change, and nature can regenerate. How will we change, and then what will we do? We need will or intention to actually act on what we've learned and according to our capacity. Will points us in the right direction and fuels our actions; it's the difference between worrying in circles and being mired in our habits or answering the call to action of our planet and getting shit done. Finally, action—where I hope the conversation is headed. When I tell people I'm writing a book about the climate crisis, nearly everyone responds with some version of "Well, yes, but what can I do?" The last part of this book aims to help us all answer that question.

When something isn't getting done—in this case, doing enough about the climate crisis—the question then is, at which of these four points are we falling short? Knowledge? Capacity? Will? Action? What is stopping us from rising to this occasion? This is what I want to talk about, and I have been surprised and encouraged—including surprised at feeling encouraged—by many of the answers I've gathered here.

For example, I had heard of climate feedback loops, but I didn't know they could go in the other direction, from damage to healing, to help us *cool* the planet. Nature, including human nature, has more capacity to heal and flourish than I previously gave it credit.

For example, when I talk about the climate crisis with other sensible, caring humans, I feel better, not worse.

For example, climate grief and anxiety are perfectly appropriate responses. We *should* be heartbroken at seeing, as Greta said, "exactly what is happening." Yet there is beauty in grief because it comes from love. Our grief can be a wellspring of will to act because it reminds us of what we love, what we have not yet lost and want to protect.

For example, what each of us does matters more than what we think, and what we do together means everything. In a way, the problem and the solution are the same: interdependence.

For example, there is possibility in uncertainty.

For example, this conversation is not about what we must give up but about everything we stand to gain.

For example, the solution to the crisis we face isn't wholly technological or scientific but also lies within our minds and hearts and in connection with one another. The potential at this time, on this Earth that is our home—"our only home," as the Dalai Lama likes to say—is enormous; potential for a world so beautiful that we have only begun to imagine it. If we rise to this occasion, we could realize a world filled with equity, beauty, abundance, and kindness.

IT'S NOT NEW that the Dalai Lama and Greta Thunberg are interested in the causes of global warming and the future of Earth, and perhaps you've listened to or read their thoughts on these topics before. But their voices together, and with the scientists, uncover more layers and unexpected insights than any one of them alone. Just as the answer to a complex riddle is often "in between the lines," we may be surprised to find in conversation with them and with one another the knowledge, capacity, will, and action to start living now for a future we can love.

Nor did their conversation have all the answers. It was a starting point, a beacon, a call to action, and, I hope, a moving symbol of people—very different people—coming together and talking about the crisis we are in. The truth is our teachers can show us where to go, but they cannot tell us what we'll find when we get there or along the way. Many people around the world seem to understand this. Despite the insurrection in Washington, DC, that had happened on January 6, 2021, three days before our event, and was still overwhelming the news and dominating much of the world's attention, nearly a million people tuned in to watch the Conversation live, and many tens of thousands more have watched the recording afterward. It was simultaneously translated into thirteen languages and reached people on every continent. By further broadening the conversation to include millions of people who have not felt personally involved with climate activism or have been blocked by confusion, paralysis, shame, or fear, I hope we can change how we talk about the climate crisis and, in so doing, realize the causes and conditions to change course. Together with some of the best of the best of social, environmental, and cognitive scientists, as well as leading teachers of ancient wisdom traditions, I want to talk about a path to awaken the bodhisattva within all of us: from the knowledge we need, through the capacity we have and the will we can ignite, to the action that will make a difference.

Greta wrote a book with her family, *Our House Is on Fire: Scenes of a Family and Planet in Crisis*, which I have given to all my close friends and family and cannot recommend enough, not only as part of the climate conversation but also as a moving family story and beautiful piece of writing. Greta's speeches have been collected and published. We will encounter her words throughout this book, and I am grateful for them. But my intention is not merely to get more people facing Greta but rather to get more adults facing the climate crisis. I wrote this book to help take a load off the young people, not to ask them to do something for me. As the ones with the most at stake in the future, they already bear too heavy a burden. One reason that Greta and the youth movement get through to people is that they tell the most compelling story yet about the climate crisis, a story Yuval Noah Harari sums up as the old sacrificing the young "on the altar of [their] greed and irresponsibility."[12] I want to say to my fellow older people, on younger people's behalf, "Yeah, stop doing that." As the Dalai Lama acknowledged to Greta, "I think anyway, I may say, our generation created a lot of problems." So I'm saying let's listen to the young people. Let's catch up, smarten up, and step up. Also know that all the author proceeds of this book go to the Mind & Life Institute, a nonprofit organization dedicated to the flourishing of all: the whole planet—humans and the natural world. "If everyone knew how serious the situation is and how little is actually being done," says Greta, "everyone would come and sit down beside us."[13] This book and my life since her conversation with the Dalai Lama is me sitting down beside them and inviting others to do likewise.

Do you know the children's story about a bunch of woodland creatures who seek shelter from the rain under a mushroom? More and more of them keep coming and seemingly magically they find room. It turns out it isn't magic; it's what mushrooms do in the rain—they get bigger. I wrote this book with the hope of growing the conversation about climate change, not just bigger but deeper—and not as a set of sacrifices or demands but as a refuge. This book is not about me. I'm just here to say, Come on in under the mushroom. We are gathering—you, Greta, the Dalai Lama, the scientists, the activists; the young people, the boomers, the woodland creatures, and everyone in between—and things are going to grow and change with us. This conversation is going to change us if we let it. As Greta says, "Spread that knowledge, spread that awareness to others."

As the Dalai Lama said in conversation with Greta, "Past is past. The future depends on younger generations." As it happens, I'm writing the last lines of this introduction on his eighty-seventh birthday. He's still working to spread this knowledge. "Younger generations" is the rest of us.

PART ONE

Knowledge

THE HUMAN BRAIN is something very special, remarkable. Yet to judge from our world, we human beings are also the greatest troublemakers. Other animals, you see their daily life: eat, sleep, sex; but we are not so simple. We have much desire, worries and concerns, urges and cravings and feelings. And too much sense of "we" and "they." I think among the different species of mammals on this planet, we human beings create a lot of good things, but at the same time, we create a lot of problems.

—THE DALAI LAMA

WE NEED TO TELL PEOPLE about what's happening right now because we are, to a large extent, unaware of what's happening. Most people I know, I encountered, haven't even heard of feedback loops or tipping points, chain reactions and so on, but they are so crucial to understanding how the world works.

—GRETA THUNBERG

1

THE SCIENCE

Why the Ice, Wind, Clouds, and Trees Matter

"WELL, THE SPEECH IS VERY DRAMATIC," Greta says wryly, by way of introducing a clip we are about to watch together. She is referring to her unforgettable, righteous, how-*dare*-you address to the United Nations (UN) in 2019. "But I thought that, at least my experience is that right now there is a huge lack of awareness and that the science isn't being enough discussed and brought into conversation."

Again and again, Greta Thunberg (pronounced grAY-tah tOOn-bay-ree, by the way[1]) has implored us not to take her word for it. She brings the latest Intergovernmental Panel on Climate Change (IPCC) report to stand as her testimony before the United States Congress. She says, "Listen to the scientists." Thanks to the movement that coalesced and took off in the years since her first school strike, in 2018, I had been getting a sense that more people were listening—more, but perhaps not enough. Because that morning in January 2021 with the Dalai Lama, more than a year after that UN speech, she still wanted to talk about a lack of awareness. In one conversation, she came back to the problem five times.

"I think we are desperate for a raise in spreading of awareness, and we need to tell people about what's happening right now because we are, to a large extent, unaware of what's happening," she said. Specifically, "Most people I know, I've encountered, haven't even heard of feedback loops or tipping points."

How much knowledge should a regular person, someone who's not a climate scientist, have? How much is helpful? How much can we stand? I wrestle with these questions. I don't know the answers for sure, but I can tell you that not being alone with the questions has made a big difference for my climate anxiety. It's paradoxical on the face of it, but the more time I spend listening to the scientists—especially when we're talking to one another as caring people and I get a felt sense of fellow humans living through something together, wondering and scratching our heads and trying to do our best—the less I feel like I'm in a nightmarish free fall into some dark, unfathomable abyss. The climate scientists you will meet in this chapter, who were part of the Conversation with Greta and His Holiness or in the Climate Emergency: Feedback Loops series, the films the conversation launched, are hardly kumbaya types. When they speak, they sound gentle and kind, but they are also extraordinarily practical, rational people, some of their fields' best scientific minds. This conversation is not a therapy session, yet I have a feeling when I'm talking to them that I am being held.

And I can tell you that I'm especially interested in knowledge that, without a whiff of Pollyanna, nonetheless gives reason to hope and points to action. Knowing about climate feedback loops is this kind of knowledge, and so, taking our cue from Greta, they will be the subject of this chapter's conversation. Have you heard of climate feedback loops? Scientists have identified dozens of them already in motion; in this chapter we'll talk about five of them and why they matter. These feedback loops are a revelation to me because they seem to contain the whole truth of the climate crisis even if you consider just one—not the whole story but the whole deeper truth of where we find ourselves.

Feedback loops are both the bad news and the good news. They're the emergency and the possibility. Knowing more about them, we can see why we can't keep going the way we've been going, and we can also see how nature itself will help us fight the climate crisis if we let her. Scientists say nature, if we help turn these feedback loops in the opposite direction, will be our most powerful ally. Let's listen.

THE EMERGENCY

We know the Earth is warming. We know the burning of fossil fuels like oil, coal, and natural gas are filling the atmosphere with heat-trapping gases—such as carbon dioxide, methane, nitrous oxide—at levels humans have never seen before. And as the world debates how much warming the planet can take above pre-industrial levels—1.5 degrees Celsius? 2 degrees Celsius? 2.5?—the climate crisis escalates.

But not many of us know that it's more than our emissions heating the globe. The rising temperatures from the effects of greenhouse gases are setting in motion Earth's own natural warming mechanisms that then feed upon themselves.

George Woodwell has been sounding the alarm about these warming mechanisms for the past fifty years. George is a climate scientist so distinguished that the research center he founded on Cape Cod in 1985, Woods Hole, now calls itself the Woodwell Climate Research Center in his honor. In an article published in *Scientific American* in 1989, he wrote that warming caused by human activity, "rapid now, may become even more rapid as a result of the warming itself."[2] Today, George has dropped the "may." "The problems are that the world is becoming too hot for the present distribution of people, agriculture, human welfare and human interests," he says, "and it's getting worse."[3]

Thirty years after George's *Scientific American* article, sixteen-year-old Greta, addressing the United Nations, repeated his warning: "The popular idea of cutting our emissions in half in ten years only gives us a 50 percent chance of staying below 1.5°C and the risk of setting off irreversible chain reactions beyond human control. Fifty percent may be acceptable to you. But since those numbers don't include tipping points, most feedback loops, additional warming hidden by toxic air pollution, nor the aspect of equity, then a 50 per cent risk is simply not acceptable to us, we who have to live with the consequences."[4]

But what exactly are feedback loops and tipping points leading to irreversible chain reactions? I hadn't heard about them until recently. Now that I have, I can't believe they're not a part of everyday conversation.

Kerry Emanuel offers the familiar example of audio feedback, "where if you put a microphone too close to a speaker, you get this terrible high-pitched screaming. And that happens because the sound comes out of the speaker and it goes back into the microphone. That's called a positive feedback because it amplifies the loop." Kerry is a meteorologist, a prominent climate scientist at MIT, and a sweetly affable guy who wrote a little book in 2012 called *What We Know about Climate Change*. (An updated edition came out in 2018.) I remind myself that when you're talking to scientists, "positive" and "negative" don't refer to the goodness or badness of what results. By "positive," Kerry means any feedback system that is self-reinforcing. Virtuous circles and vicious cycles are both positive feedback loops.

When it comes to climate, emissions from the fossil fuels we use are the input, adding greenhouse gases to the atmosphere, raising Earth's temperature, and setting in motion self-perpetuating warming loops of melting ice, snow, and permafrost; a meandering jet stream; heat-trapping clouds; and burning and dying forests, all of which contribute to further warming—or warming as a result of the warming itself. That increasing screeching noise Kerry was talking about may strike you as an apt analogy for the damage that human-caused feedback loops are wreaking on the planet.

The Melting Mirror Loop

Climate geeks talk about albedo (rhymes with Speedo). I'm not sure if this term will make it into everyday conversation—or if it should, since we have other words that people readily understand, like *mirror, reflectivity,* and *ice.* What the scientists mean when they say "albedo" is the reflectivity of a surface. It could be any surface, but they're particularly concerned with the Arctic. At Earth's poles, I learn, snow and ice reflect up to 85 percent of the sun's rays away from the surface and back into space, helping to keep the planet from becoming too hot. But over the past few decades, this natural mirror has begun to break down as fossil-fuel emissions raise temperatures and melt snow and ice cover. As the planet loses its ability to reflect sunlight, a dangerous warming feedback loop is triggered. The most alarming change is happening in the far North, where the temperature rise is causing the snow cover and sea ice to rapidly disappear. Scientists can keep calling this the "albedo effect," but I think of it as the "melting mirror" loop, like something in a painting by Salvador Dali, which seems appropriate for these not-normal times.

The sea ice geophysicist Donald Perovich has documented big changes in the Arctic. "There's always been this annual cycle, the ice grows usually for nine or ten months of the year and then melts for a couple of months," Don explains. He has the kind of eyebrows that if he wiggled them, they would make you laugh, but this is not eyebrow-wiggling subject matter, and his face remains steady, though not unpleasant. "What's changing now is the timing. The melting is starting earlier, the freezing is starting later. We have much less coverage every month of the year, particularly at the end of summer."

Here's why all the fuss about the Arctic, why even if you haven't been paying much attention you know the climate crisis has something to do with polar bears. Global warming from greenhouse gas emissions from human activity is increasing the temperature in the Arctic two to three times faster than the rest of the planet, and this affects *everything*. (Climate change communications experts now recognize that we need different imagery to spread the message; that it is not just a polar bear problem but a people problem.)

For those of us who've never flown over the Arctic, Don shares a bird's-eye view. "Say it's April and we're flying above the Arctic, and we look down at the sea ice cover. It's covered by snow, it's bright, and it's white. Now summer comes, that snow melts, you get more open ocean. It's absorbing much more heat. Instead of reflecting 85 percent, it's absorbing 90 percent. And so it's replacing one of the best natural reflectors, snow, with one of the worst, the open ocean." The ocean is dark. Don says he is optimistic by nature and that he becomes "more optimistic when I see so many people that realize there's a problem." But this feedback loop, in the direction it's going, is dark both literally and figuratively.

"As those darker waters warm, they emit carbon dioxide and water vapor, warming things further," George Woodwell adds. He says those last three words so heavily they should almost have periods: Warming. Things. Further. "So, there are several aspects to this feedback in the Arctic problem that are truly frightening." Apparently the volume of sea ice has decreased by 75 percent in the past forty years, and studies suggest that around a quarter of global warming is caused by the loss of this sea ice. But if you factor in the melting of snow cover on the surrounding land, together they account for an approximately 40 percent loss in the planet's reflectivity, a number that with further warming will keep growing.

"Our climate model projections suggest that we will lose the Arctic sea ice cover in the summer months altogether by the end of this century," says Marika Holland, a climate modeler from the National Center for Atmospheric Research in Boulder. "If we continue to increase greenhouse gases in the atmosphere through the burning of fossil fuels," Marika continues, "we ultimately will get to a state where we lose the winter sea ice as well."

To put this in perspective, it helps to know that ice has covered the Arctic Ocean for more than two and a half million years. Marika also helps me understand why what's happening in the Arctic isn't only a problem for polar bears, patiently explaining that "the Arctic plays a very central role in Earth's climate. Even if you just lose sea ice cover in the Arctic, the tropics will feel that enhanced warming." In other words, what happens in the Arctic doesn't stay in the Arctic. The warming Arctic air mixes into the global atmosphere and raises global temperatures, exacerbating problems the climate crisis is already causing—crops suffering, food prices going up, wet areas becoming wetter, dry areas becoming drier. Marika says, "The models predict if we continue on the path we're on, the Arctic will experience very dramatic changes and those changes will reverberate throughout the system, the human system, the biological system, the socioeconomic system." Scientists talk about the "New Arctic" now because it's suddenly so different from what it has been for millions of years.

As the climate warms, it kicks in another feedback loop, set in motion by the melting of massive glacial ice sheets. "Loss of ice over the Antarctic continent"—miles-thick sheets that have been accumulating on land for over forty million years, Marika says—"doesn't have as much of an albedo feedback because it's so thick" (too thick to melt entirely). "But as that ice enters the ocean, it causes sea level to rise." The ice sheet that covers Greenland is also melting into the sea. As the oceans rise, the higher, warmer water melts more land ice, raising sea levels further and melting even more ice in a vicious cycle.

The scientists tell us that if both Greenland and Antarctica's glacial ice shelves were to melt, sea levels could potentially rise by more than a hundred feet. A friend in New York tells me that the East River Park near where she lives in the East Village is currently undergoing a massive renovation to raise it nine feet. The park follows the river in a long narrow band for nearly ten miles, and during Hurricane Sandy in 2012, both the park and its adjacent neighborhoods were flooded. The raised park, along with improved drainage, is supposed to protect against flooding when the next sea surge comes (the East River is technically an estuary). The renovation is projected to take three or four years, during which time the park and its beloved bike paths, river views, summer barbecue parties, and dog parades will be mostly closed. How many years of protection from rising seas will those neighborhoods get, I wonder, from *nine feet*?

The scientists tell us that if we continue with "business as usual," to use a favorite phrase of Greta's,[5] the warming in the Arctic will cause the feedback loops at both poles to spin out of control.

The Spoiling Food Loop

"It's like having a chicken in your freezer and then you take the chicken out, you put it on the counter, and it starts to thaw," Sue Natali says. She continues with the analogy, "And then you go away for the weekend, you forgot about the chicken that's on the counter, and you come back and the house smells and the chicken's decomposed. That's what happens to the carbon that's in permafrost." Sue directs the Arctic program at Woodwell Climate Research Center and has devoted her career to analyzing Arctic soil and why it matters to all of us. Her freezers contain actual core samples—dense cylinders of earth with the girth of large salamis that she and her fellow field researchers have collected each year at sites they return to in Alaska and Siberia. Her chicken comparison is closer to home and helps me understand what this soil has to do with climate.

Permafrost refers to a layer of soil that stays frozen all year round. In the Northern Hemisphere, Sue explains, nearly one-quarter of the land is permafrost. Extending thousands of feet below the surface, this soil contains carbon-rich plant and animal remains suspended in a perpetually frozen state—that's the "chicken" part. With human activity warming the Arctic (recall, two to three times faster than the rest of the planet), this permafrost is starting to thaw and the carbon in it, like the chicken left for days at room temperature, is starting to rot. All those animal and plant remains in the soil are "fuel for microbes," Sue says, referring to the microscopic animals that have been frozen for up to tens of thousands of years, which are now waking up and feeding on the thawing carbon. "As they're breaking it down and using that fuel, they're releasing greenhouse gases, carbon dioxide and methane, into the atmosphere." With the chicken out of the freezer, so to speak, the microbes can't help themselves; as long as the soil is above freezing temperature, they will feast on the carbon. In their own way, the microbes are like us, burning fossil fuels like crazy. But how much damage can some microscopic animals do?

Apparently a lot. According to scientists, all the microbes on Earth weigh probably fifty times more than all the other animals on Earth. Combine that with the fact that permafrost contains *a lot* of carbon, and you get a feedback loop of warming, thawing, releasing greenhouse gases from the soil, more warming, and more thawing at a scale that is already having global impacts.

Sue offers some numbers that help us fathom the magnitude of this feedback loop: "The amount of carbon that could be released from thawing permafrost by the end of the century has been estimated to be up to 150 billion tons of carbon," she says.

"So, to put that in context, the United States is currently the second largest greenhouse-gas-emitting country in the world. If we took our current US emissions and added them up through the year 2100, this is on par with the amount of carbon that might be released from thawing permafrost."

Last summer, while working in her usual field location in Alaska, Sue says she witnessed a remarkable acceleration of permafrost melting. "First of all, it was very, very warm. It was ninety degrees Fahrenheit in the tundra," she recalls. "There were places where we walked, where my foot fell into the ground because there was no longer any ground structure because the permafrost was thawing. I've never seen change happening that quickly from one year to the next."

The thawing permafrost not only impacts the climate through the release of greenhouse gases; it is transforming landscapes, as Sue has seen in Duvanny Yar in far-northern Russia. "I had never seen permafrost thaw and ground collapse of that magnitude. I remember driving up to it on the boat and it was like, wow, this huge cliff, like, many, many stories high." Fine roots in the soil that held the ground together have been frozen for forty thousand years, Sue says, but once they thaw, they decompose within a year. When people live on or near these collapsing landscapes where the permafrost is melting—for example on the Tibetan Plateau—obviously it's a local problem. And when the permafrost feedback loop continues in the direction it is going, it affects everyone on the planet. However, as with all these feedback loops, the reverse is also true: since greenhouse gases mix throughout the global atmosphere, action everywhere in the world can reduce permafrost thaw—and reduced permafrost thaw would decrease warming everywhere in the world.

I am shocked to hear from Sue that permafrost feedback has not been accounted for in the global carbon budgets that have been used to determine how much and how fast we humans need to reduce our greenhouse gas emissions to limit warming. Unlike the microbes, we have choices. "It's critical that decision-makers are aware of and account for the permafrost carbon feedback," she says, "in order to keep our climate under control and to restore the planet."

Greta has a question: "Could you clarify in what way these effects, these feedback loops, are not included in, for example, the global carbon budgets and why is that?" Sue explains, for example, in her area of research, that the models that inform the latest reports that advise these global carbon budgets didn't include permafrost because these models are built for the planet, not specifically for the Arctic. More generally, she allows that the scientists are well aware of such omissions and are actively trying to get these processes into the models, "but sometimes science acts a little bit slower than it needs to, when we think about policy, when we think about a climate emergency. Even if the models aren't perfect, when we think about managing for risk, we need to act. Maybe we don't know the specific number, maybe we know a range of numbers, but those numbers are good enough for us to communicate what these risks are." The scientists who work with these models are quick to say, when they present their findings, that it's actually worse than that, because of feedback loops.

WHEN PESSIMISM IS NOT AN OPTION

His Holiness the Dalai Lama

Today, we humans have no excuse for not knowing about the climate crisis. Study after study by scientists, as well as the dramatic weather-related changes we are witnessing more and more—

extensive flooding, draught, heat waves, and powerful hurricanes—have laid it bare for us. They tell us that we are facing a future where the very survival of humanity is threatened if we do not act immediately. Now, when we are presented with such a stark existential threat, and when we appreciate the enormity of the task ahead to counter it, quite naturally we might feel afraid and even overwhelmed. This fear can lead us either to pessimism and loss of hope or it could equally lead to greater determination and commitment to doing whatever we can. Personally I have never been drawn to pessimism as it is a form of defeatism—it means giving up before trying. In the face of climate crisis, I don't think we have the luxury of giving up. We owe it to future generations, and to ourselves as well, the knowledge that at least we tried.

We might ask, how can we maintain optimism even as we become more aware and educate ourselves? Here are a few things I do to steady and uplift myself:

Find a way to adopt a wider perspective. Sometimes our sense of overwhelm and powerlessness comes from having too close-up a view of the problem. However, when we situate the problem within a larger context, say, in the case of climate, we can see that the very forces—interdependence, causation, and dynamic effect—that create the negative spiral toward global warming, these same forces can help create a virtuous cycle once we begin to make positive changes.

Take inspiration and encouragement from past successes. If we pay attention, we will find many local and global stories where human efforts have made real difference, be it regreening, developing clean energy, reducing our use of plastic, et cetera. Reflecting upon these successes can lift our hope and give us more determination.

Steadfastly maintain our hope and courage. Finding hope and maintaining it is crucial if we are to have any success in meeting our individual and collective challenges. Hope allows us to envision a better future, and it brings with it a positive energy, which is crucial for our motivation. With hope we can have the courage to care and the courage to act. In any case, giving up is not an option we have with respect to the climate crisis.

Do not lose hope in humanity. Although we humans are the ones that have created today's climate crisis, we must find a way out, to save the planet we share with so many other species. The trend toward stricter environmental regulations, greater willingness to come together in collective forums such as international climate summits, technological innovations toward green energy, and growing recognition for the critical challenges of population growth and lower consumption—all of these give me hope. In any case, the fact remains that if there is going to be any solution to the problem of climate change, it's humans who must come up with it. No other species can.

Bringing these aspects to my mind helps me to maintain my hope, courage, and commitment to be part of the solution we all seek.

The Meandering Jet Stream Loop

The jet stream is one of those things that we hear about all the time, and we know it affects the weather. However, if I had to explain it to someone, I would be of little to no help. Thankfully Jennifer Francis knows the jet stream inside and out. Jennifer studies how increased greenhouse gases affect the atmosphere. A senior researcher at Woodwell Climate Research Center, she is particularly interested in the jet stream. She describes the jet stream almost poetically, as a "river of wind high over our heads up where the jets fly." The jet stream encircles the Northern Hemisphere, Jennifer says, and "is responsible for creating pretty much all the weather that we experience in this part of the world." Then she guides us on a little jet stream visualization.

"Imagine a layer of air extending from the warm South to the cold North. Warm air expands, so the layer over the South rises higher than the air over the North. So then imagine you're sitting up on top of this layer of air in the South looking north—it would appear to slope downward. Because of gravity, the warm air higher up flows downhill just like water flows down a mountain." I think of her "river of wind" description. "This downward movement creates a south-to-north wind. But because the Earth is spinning, this wind gets turned to the east and becomes a west-to-east flow of wind. That's the jet stream."

The greater the temperature difference between the north and south air masses, the faster, stronger, and straighter the jet stream winds blow. Historically, Arctic air has been much colder than air to the south, keeping the jet stream fairly straight, with relatively small north-south meanders. But with the Arctic warming faster (say it with me: *two to three times faster*) than the rest of the globe, that temperature differential has decreased. This weakens the jet stream winds, causing them to take larger swings north and south, which in turn impacts the weather.

Again with the Arctic. I knew the Arctic figured in conversations about climate change, but I didn't realize there were so many reasons why. Yes, Jennifer says, the jet stream is yet another feedback loop involving the Arctic, because the more extreme southerly meanders are pulling more heat from south to north: "We're warming the Arctic, we're reducing the winds of the jet stream, we're seeing it take these bigger north-south swings, which then transfers even more heat from the South to the North into the Arctic, which makes it even warmer, which weakens the winds more, and it sets up this vicious cycle." Jennifer says on the ground we experience this as our weather getting "stuck" in one pattern—hot, cold, rainy, or dry—for a long time. "The wet places are going to tend to get wetter," Jennifer explains. "On the other hand, places that are already dry, they're also going to see more evaporation because of the air being warmer." These extremes in every direction are what the climate scientist Katharine Hayhoe famously calls "global weirding," to preempt anyone trying to use the existence of a cold snap or a snowball as evidence against global warming.

Both the current multiyear drought in the western United States and the increase in wildfires are connected to a larger north-south swing in the jet stream. Both, not very intuitively, have something to do with the Arctic. The scientists say we must turn this feedback loop around: reduce greenhouse gas emissions, cool the Arctic, straighten the jet stream, further cool the Arctic. They also say we're already living in a world where extreme weather events are the norm, not the exception, and that even if we could staunch emissions of greenhouse gases today, these weather patterns may continue for a long time.

Years ago, I came across an article in a popular magazine about the effects of climate change and, rare for me at the time, made myself read it.[6] In addition to considering the very real possibility of a major coastal city being underwater, the author, Andrew Rice, was interested in a kind of philosophical dilemma that comes with the climate crisis, having to do with the mitigation-versus-adaptation debate. His point was that it's difficult for us to hold both these priorities at the same time. Adaptation and mitigation can seem to contradict each other, making it hard to talk about the former without sending a confusing message: because an adaptation mindset assumes that we have failed to prevent climate change, it suggests mitigation is futile. Listening to

Jennifer talk about the jet stream, though, I feel shaken out of such black-and-white thinking. Clearly we must adapt to weather patterns that are already here to stay for some time; and we must do everything we can to reverse climate feedback loops, to not make things worse and eventually, hopefully, to make them better. Not either-or but both-and.

The Blanketing Clouds Loop

To paraphrase Joni Mitchell, we really don't know clouds. The National Medal of Science recipient Warren Washington began creating computer models in the 1960s to predict the future of atmospheric warming and the role feedback loops play. Today in his eighties, after a long and august career at the National Center for Atmospheric Research, he humbly admits that "clouds are very complicated."

"We still don't have good mechanisms," Warren says, "as good as we would like, for cloud feedbacks." But we know something. We know clouds are formed by water vapor, a naturally occurring gas created when water evaporates from lakes and oceans, and we know that water vapor is also a heat-trapping gas. In fact, water vapor accounts for about 60 percent of all global warming caused by heat-trapping gases in the atmosphere. Sixty percent? Really? Water vapor feedback, scientists say, amplifies global warming from human activity between two and three times.

I think about the last time I lay on my back and looked up at a sky half-full of streaky, puffy, lacy, seemingly endless variations of clouds. I think about the relief on a very hot day when the sun goes behind a cloud. I think about images of Earth from space, the big blue marble aswirl with clouds. They appear so natural, if not benign—how did they get caught up in our climate crisis?

"Water vapor is just water in a gaseous form," Jennifer Francis says, taking us back to kitchen basics. "When you take a pot of water and put it on your stove and boil it, you see steam, which is still in the liquid form, but then it disappears, and it goes into the atmosphere and it's completely invisible. The same kind of thing is happening in the climate system where, as we warm the air and we warm the oceans, more evaporation is occurring from the oceans and putting more water vapor in the atmosphere."

While some of this water vapor stays in the atmosphere and traps heat in gaseous form, some of it cools and condenses, forming clouds, which can both heat and cool the planet. Clouds can lower the temperature because their white color reflects sunlight back into space, cooling the Earth. Clouds in this capacity are why we think of cloudy days as being cooler than sunny ones, other things being equal. But conversely, clouds can trap heat below them, heating Earth's surface like a blanket, as when it's warmer on a cloudy night than a clear one. So, what's the net effect for the planet? As Warren said, it's complicated, but in the final analysis, scientists have concluded that clouds have a warming effect. And as the climate warms, the oceans are also heating up, causing increased evaporation, creating even more water vapor, trapping more heat, leading to more evaporation, in another amplifying loop.

It's this combination of increased water vapor and the warming of the oceans that's driving hurricanes, scientists say. Hurricanes aren't part of the water vapor feedback loop, but they're profoundly affected by it, and climate models predict more frequent stronger hurricanes in the future. "More than thirty years ago we predicted that global warming would result in more intense storms," meteorologist Kerry Emanuel concurs. "And we're beginning to see that, even in places like Florida and the Bahamas, which are adapted to hurricanes, they're adapted up to a point, but when we begin to see stronger storms, like Dorian in 2019, that adaptation doesn't mean anything."

I think he means "doesn't mean anything" in the sense that adding nine vertical feet to your seawall doesn't mean anything if the oceans rise by fourteen or a hundred.

The Disappearing Trees Loop

My friend Elissa Epel and I have a thing about trees. Elissa is a professor of psychiatry at University of California San Francisco and an all-around radiant being. We live on opposite coasts, but since she is also deeply involved with Mind & Life—she has been faculty and served on a number of committees—we get together regularly enough, and whenever we're together we make a point of going for a walk with the intention of finding a special tree. It happened spontaneously the first time, at a big conference in San Diego, but over the years this thing has grown into a ritual, a kind of walking meditation on roots and trunks and light and leaves until we find a tree that seems to speak. Without fail, we both know when we have found "our tree." We stop to appreciate it from ten or twenty feet, taking in the grandeur, sway, and symmetry; then we approach, slowly, until we're looking up from under the canopy. We stand there for a while. We close our eyes and feel ourselves enveloped by the shade. We look up and see the sky between the leaves. We hug the tree, hug each other, offer some gratitude to this life-giving being, and always leave with some photo of our tree; sometimes we're both in the photo—a tree selfie. I have one of Elissa closely inspecting the leaves of a fifty-foot ginkgo. I remember how its lower branches hung down like strong, tired arms, low enough to touch, and we marveled at the Asian fan-shaped leaves.

"Tree hugger" is an anti-environmentalist slur, a sneer, as though loving trees is weird, creepy, and fringe. But the more I listen to scientists talk about trees, the more reasonable loving them seems. I can't help but think that more people would love them and want to protect them if more people knew what they do for this planet where we coexist with trees. We need to know that we need trees, and because we need them, we need to stop killing them as indiscriminately and rapaciously as we have been. Is anyone an "anti-environmentalist" anymore? Really? We should all be environmentalists, it seems to me. Where else would we live? As the Dalai Lama likes to remind us, this planet is our only home.

Unscientifically speaking, trees are *miracles*. In scientific terms, *photosynthesis*—let's remember how amazing it is. Trees feed themselves with sunlight, water, and carbon dioxide. I've heard technology optimists talk about carbon capturing as the technological savior that would, if we figured out how to do it at scale, effectively let us reach up and pull enough carbon out of the air. Wouldn't that be amazing! And planet saving! But trees (and other green plants) have already figured out how to do this amazing thing. It's what they do all day, every day, by simply being. Trees reach out with their leaves and branches and literally pull carbon dioxide out of the air, store it safely away in themselves—in their branches, trunks, leaves, roots, and soils—and release oxygen instead. And the Earth's forests already know how to do this at scale. Trees cool the Earth if we let them. This is the incredibly good news contained in the forest feedback loop. In fact, scientists tell us, every year terrestrial ecosystems remove about 30 percent of fossil-fuel emissions, and forests are responsible for most of that.

But here's the bad news, with the direction we're currently headed: That percentage is decreasing as our greenhouse gas emissions increase, steadily raising Earth's temperature and threatening forests' ability to offset the warming. As trees burn or decay, the carbon they've locked away during their lifetime—what scientists call a "carbon sink"—is released back into the air. "We have warmed the Earth by a full degree and a little more," as George Woodwell points out, "and forests are suffering increased hazards of fire as they get warm and dry, and increased hazards of disease as they become vulnerable to insects—and are dying as a result."

As more and more of the Earth's forests are destroyed for agriculture and industry, there won't be enough trees left to pull enough carbon out of the air to keep up with the warming. So the temperature rises still higher; the climate becomes hotter and drier; and more trees fall prey to insects, drought, and fire.

"It's entirely possible we reach a point where we're killing off forests much more rapidly than carbon can be fixed by forests," George says. "The net result, then, is to produce a feedback that's lethal."

When it comes to global warming, scientists say that three major forests matter most: the tropical, the boreal, and the temperate. Michael (Mike) Coe is the director of the Tropics Program at Woodwell. Gathering data in the field in sun protection and galoshes, Mike studies how deforestation in the Amazon rain forest affects the local climate and environment. He estimates that tropical rain forests account for about 15 to 20 percent of all the terrestrial carbon sink. The Amazon, he says, is responsible for half of that, "so, we're talking about a significant fraction of our annual emissions being taken up by this forest." Spanning more than two million square miles across nine countries, this tropical forest has been storing carbon for millennia, yet it's dangerously close to releasing more carbon than it absorbs.

Why? Because in the last fifty years, nearly one-fifth of the Amazon has been lost, mostly to slash-and-burn land clearing, fires, insects, and tree dieback. This not only releases the carbon stored in the Amazon, scientists say, but it also jeopardizes an important cooling function of the forest. During transpiration, Mike explains, roots pull water out of the ground and release it as water vapor through tiny holes in the tree's leaves, creating a cooling effect on the surrounding air. In the Amazon, transpiration can cool the region by as much as ten degrees Fahrenheit.

But Mike wants us to know, "When we lose trees in the Amazon, when we cut them down, what we're doing is we're shutting off that transpiration. So what you get is a drier climate, and the more you deforest, the drier it gets." Another vicious cycle. "During extreme droughts, a huge amount of the forest burns," he explains. "That turns the forest that year from a net sink to a net source of carbon. We can do the math. If that happens five times a decade, this forest is no longer a carbon sink but a source."

Today, the scientists say, tropical forests are absorbing one-third less carbon than they did in the 1990s. They predict that with the loss of so many trees, the Amazon could shift to emitting more carbon than it stores as soon as the next decade. "You can imagine that the climate's changing a little bit, and everything looks just fine," Mike says, and I've learned enough now to hear the "but" coming. "But then one day, it changes just enough that the whole system flips over." This is what Greta means by a tipping point.

The next major forest scientists say is at risk of tipping from a carbon sink to a carbon source is the boreal, encircling the North Pole through Siberia and North America. The largest forested region in the world, this vast coniferous expanse stores an estimated two-thirds of all forest carbon, most of it locked away in frozen plant and animal remains deep in the ground, but that's changing. As with tropical forests, the warmer, drier climate is making these trees more vulnerable to disease, insects, and fire.

"Wildfires are getting worse across the boreal zone," says Brendan Rogers, a climate scientist at Woodwell whose work focuses on the forests of Alaska, northern Canada, and Siberia. "We're seeing more and more large fire seasons, record-breaking fire seasons, every year."

Unlike in tropical forests, the northern forest fires strip off the insulating ground cover, preventing it from building up in between the frequent burns. Without this protective layer, fires reach further and further down, burning the organic matter stored in the soil. "Seventy-five to 90 percent of all the carbon stored in these forests is underground," Brendan notes, "and that is actually a majority of the carbon that's getting released from these fires."

So the fires kick off a feedback loop triggered by warming in the boreal zone: more fires burn carbon deeper down in the soil, releasing carbon dioxide and methane into the atmosphere, heat-trapping gases that make the climate hotter and drier, leading to more wildfires, more carbon release, and on it goes.

Like the Amazon, the boreal forest is going to flip from a carbon sink to a carbon source. Scientists say they don't know exactly when, but they predict that at the current rate, it will happen by the end of this century, crossing a tipping point that the forest can no longer recover from. Mike calls the changes he's seeing "rapid . . . so rapid that the system is essentially not used to it. And the scary thing is not knowing where that tipping point is."

The temperate forest, meanwhile, makes up only one-quarter of the Earth's forests, but the scientists say that since the tropical and boreal forests are on the brink of becoming emitters of carbon, the temperate forest is our best hope. Once cleared for agriculture and lumber—"Clean-shaven like a monk's head," I recall the Dalai Lama's vivid simile—many temperate forests in Europe and North America have made comebacks in recent decades. But in the southeastern United States, apparently, old forests are being cut down by the wood pellet industry for burning, releasing decades of stored carbon back into the air.

The wood pellet industry? How many people have even heard of it? If you haven't, I recommend Sarah Miller's excellent article about it in *The New Yorker*.[7] Another word for wood pellets is the euphemistic "biomass," a source of so-called bioenergy— words the energy industry seems to like, I guess because they obscure the fact of forests being razed and burned, and they sound vaguely "green." But wood pellets are just what they sound like: trees reduced to slender bullets of compressed sawdust that ship easily, often across oceans in freighters powered by fossil fuels, to refurbished coal plants where they're burned instead of coal. On a technicality, because trees can be and are being replanted, this source of energy is officially considered "renewable," but there is one thing about forests the scientists say we absolutely need to know: when it comes to offsetting global warming, old and young forests are not equal.

Beverly Law measures the exchange of both carbon dioxide and water between our forests and our atmosphere. She works at two long-term testing sites that are part of a network of more than five hundred around the world, and she has been taking these measurements for the past twenty-five years. She affirms what we can also intuit when we picture the massive, towering trees of an old-growth forest, with its thick carpet of rich soil, compared to the same area of newly planted saplings: new growth doesn't compare to old growth when it comes to carbon storage. As the Canadian writer Charlotte Gill describes in her memoir about tree planting, "You don't need a PhD to note the difference between a virgin forest and a recycled one. The ground here is stones embedded in sand, covered over with crusts of sun-dried moss. Digging into it with my shovel is like working a spoon down into a jarful of teeth. I scrape handfuls of dirt together and shove them around the stems. Deep rainforest replaced with low-fat soil, a trompe l'oeil. A forest-looking forest."[8]

Then consider how long it takes a tree to grow compared to how little time we have before global warming reaches 1.5 or 2 degrees Celsius. The biggest and oldest trees are hundreds and even thousands of years old, whereas scientists say staying under 1.5 degrees Celsius depends on what we do by 2030. So we cannot simply rely on replanting; we must save enough old growth. We must respect how long it takes a forest to become *old*.

"If I were to try and mitigate climate change," Beverly says, "the best strategy when you have forests that have low vulnerability to climate change in the future and they store a lot of carbon already, is keep those forests like they are, preserve them." Not only do forests like the old growth in the Pacific Northwest store carbon, but the wetter, cooler climate makes them less vulnerable to climate change over the next thirty years compared to other forests in the western United States.

But unfortunately those hardy, carbon-rich trees have been prized by loggers for centuries. When a tree is logged, one-half to two-thirds of the carbon it stores is released through decay or burning of the unused branches, leaves, and roots, as well as from the surrounding soil. If the tree is turned into wood pellets that are burned, the rest of the carbon it stores also goes back into the atmosphere. Today, 17 percent of global carbon emissions each year can be attributed to logging and burning wood pellets for "bioenergy."

Bill Moomaw says forests are the most powerful ally we have right now for removing carbon dioxide from the atmosphere. Reducing our greenhouse gas emissions is also critical, but Bill says that will not be enough. He began working on climate change more than thirty years ago as the first director of the climate program at World Resources Institute in Washington, DC; there he applied his training as a chemist to solving the depletion of stratospheric ozone (an environmental success story). Currently an emeritus professor and codirector of the Global Development and Environment Institute at Tufts University, he has been a lead author of five IPCC reports, including the one in 2007 that shared a Nobel Peace Prize. He says we need to let more trees keep doing their amazing thing, reaching up and pulling carbon out of the air—especially big old trees. We need clean energy alternatives to burning wood pellets, like solar and wind, and to "keep our carbon inheritance—that's what it is, it's an inheritance from the past—in our forests and soils." Bill sums up the importance of forests like this: "Forests are at the heart of a feedback loop that can either warm or cool our planet, and we get to determine which direction that goes. We must refreeze the Arctic to save it. To refreeze the Arctic, we must cool the Earth. To cool the Earth and stop the feedbacks, we must reduce the heat-trapping gases already in the atmosphere. Recent studies show that if we can allow more of our forests to grow, they have the potential to store twice as much carbon as they do today. This orientation toward the protection of forests, called 'pro-forestation,' has the potential to remove more carbon more rapidly from the atmosphere by avoiding harvest-related emissions. We must take advantage of the fact that larger trees accumulate and store the most carbon. We need to let trees continue to grow so they can store the carbon that we need them to store. Planting new trees is helpful but will take a lot longer to have meaningful effects."

THE POSSIBILITY

In 1990, when I was twenty-seven, I had my first conversation about environmental sustainability. I was an oncology nurse in Hanover, New Hampshire, at the time and a captive audience, since I was in a car for three hours with Donella Meadows, driving her to a medical consultation. People who were lucky enough to know Donella personally called her Dana. She was a cancer patient; a Dartmouth professor; and a brilliant, ecologically minded systems thinker (recognized with a MacArthur "Genius Grant" in 1994). I had been assigned to her care because of our mutual interest in holistic health, and we were driving to see a specialist with what we would now call an integrative practice, though at the time, "integrative medicine" didn't exist. From the passenger seat of my six-cylinder-engine vehicle, she encouraged me to question: Did I need such a big, fast car?

I'd never given a thought to fuel efficiency aside from the expense of buying gas. Until that moment I'd only thought, "Of course I would want the fastest, most comfortable car I could afford." But Dana was always thinking and talking about being aware of how we are living and how that impacts Earth and its living systems. A gifted communicator and systems theorist, she built an extraordinary career on these interests. At the time, I knew I'd never met anyone like her, but it was years later that I realized just how extraordinary she was.

In the third edition of *Limits to Growth*, a book that she wrote with two colleagues and first published in 1972, Dana says, "The depths of human ignorance are much more profound than most of us are willing to admit. This is especially so at a time when the global economy is coming together as a more integrated whole than it has ever been, when that economy is pressing against the limits of a wondrously complex planet, and when wholly new ways of thinking are called for. At this time, no one knows enough."[9] True in 2002 and still true. Dana was ahead of her time.

Dana understood feedback loops and predicted—with complex modeling informed by theory, data, and deep human intuition—the climate crisis we would face if we didn't implement drastic changes; that is, the climate crisis we are currently facing because we didn't, so here we are. "Our actions have consequences," as Greta acknowledged to the Dalai Lama in a simple expression of systems thinking and the Buddhist concept of karma.

In *Limits to Growth*, Dana continues, "The world's leaders do not know any better than anyone else how to bring about a sustainable society; most of them don't even know it's necessary to do so. A sustainability revolution requires each person to act as a learning leader at some level, from family to community to nation to world. And it requires each of us to support leaders by allowing them to admit uncertainty, conduct honest experiments, and acknowledge mistakes."[10]

Dana was one of those learning leaders, and my conversations with her in the early 1990s changed me. I never bought another car with a V-6 engine. She played a part in my decision to go to graduate school to study psychoneuroimmunology and focus on connections between people's sense of meaning in life and their immune systems; and she played a part in my continuing to think about the world in terms of sustainability. Dana beat cancer. Then she died of meningitis in 2001 when she was only fifty-nine, but I want her to stay in this conversation.

OUR ACTIONS HAVE CONSEQUENCES. Listening to the scientists in this chapter, what I'm hearing is that while human activity has kicked off natural warming loops, human insight and ingenuity could reverse their direction, turning them into cooling feedbacks instead. As the Dalai Lama said to Greta, "The human brain is something very special, remarkable. Yet to judge from our world, we human beings are also the greatest troublemakers." Let's talk about using our very special brains to get us out of trouble: to transition to clean sources of energy; to minimize waste and air travel; to ride bikes and expand public transportation; to protect and expand forests; to preserve marshes, grasslands, and all natural habitats; to use agricultural practices that store carbon instead of releasing it; to let trees and plants do their job of reaching up and taking carbon out of the air.

"Human activity" and "human ingenuity" can sound impersonal and removed, but they are us; they're what you and I do. We are all part of these feedback loops, so what are we going to do?

Most people haven't heard of climate feedback loops, Greta points out, "but they are so crucial to understanding how the world works." Well, now you have. Heard of feedback loops. The Earth is warming the Earth. And the Earth also has the power to cool the Earth, if only we would join her.

2

THE SPIRIT

The Problem with Business As Usual

THE FIRST TIME I SAW THE DALAI LAMA, I didn't. This is not a koan. Recently divorced and back in school, in Chicago, I had next to no money to spare for a ticket to see him speak. But I really wanted to go. I asked the organization that was hosting the event if I could earn a ticket some other way. "Please," I said, "I'll volunteer, I'll do anything." But they said they "don't do that." I still couldn't shake the feeling that I should be there, so on the night of the Dalai Lama's appearance at the landmark Chicago Theater, I paid more for a scalped ticket than I would have in advance and climbed to the very back of the very top of the steep-set seats to find that mine was positioned squarely behind a large structural column. I couldn't see His Holiness. I couldn't even hear.

Fast-forward twenty years later, the Dalai Lama and I were in a small room together in India and he held my hand. Thirty-something-Chicago me did not expect this, any more than Greta planned to Zoom with His Holiness when she first sat with her protest sign outside the Swedish Parliament House. Like the Buddha himself observed, nobody knows the future. Nobody ever has. But can we talk about the particular uncertainty that is coming to define our time, the loss of certainty in the ongoing-ness of life on Earth?

Recently my friend Thupten Jinpa, the Dalai Lama's longtime English translator as well as my colleague at Mind & Life, was talking with a couple of us about an ancient visual metaphor for interdependence, a central insight of Buddhism. The image, Jinpa said, is from a scripture called the *Avatamsaka Sutra*, or "Flower Garland Scripture." It sounded beautiful, and I was picturing some kind of kaleidoscopically colorful mandala unfurled from an ancient parchment scroll. Another colleague had never heard of it.

"Google it," Jinpa said. So much for romantic archaeological fantasies.

What Jinpa was referring to is known as Indra's Net. I found a translation by the Buddhist scholar Francis H. Cook describing it like this:

> Far away in the heavenly abode of the great god Indra, there is a wonderful net that has been hung by some cunning artificer in such a manner that it stretches out infinitely in all directions. In accordance with the extravagant tastes of deities, the artificer has hung a single glittering jewel in each "eye" of the net, and since the net itself is infinite in all dimensions, the jewels are infinite in number. There hang the jewels, glittering like stars of the first magnitude, a wonderful sight to behold. If we now arbitrarily select one of these jewels for inspection and look closely at it, we will discover that in its polished surface there are reflected all the other jewels in the net, infinite in number. Not only that, but each of the jewels reflected in this one jewel is also reflecting all the other jewels, so that there is an infinite reflecting process occurring.[1]

Interdependence, as depicted by Indra's Net, sounds like an especially sparkly representation of systems theory. Jinpa says the point is not so much the sparkle (another scripture gets at the same idea with dust) as the fact that everything and everyone are interconnected. There is no center; we are all in it together, depending on one another—"we" being everything in the universe, and "it" being an infinite web of cause and effect. "Of course," says Jinpa, "there is no way for the human mind to be able to grasp all the points of interconnection; for that we would need to be omniscient." Seeing that we humans, as individuals and as a species, aren't really separate and certainly aren't at the center helps us assume a wider, longer-term, and critically humbler and less selfish perspective. Jinpa says recognizing interdependence also lets us see beyond the immediate and obvious to underlying patterns, like the difference between noting today's weather and understanding our role in climate feedback loops.

Greta agrees. Feedback loops, she says in conversation with the Dalai Lama, "show how complex everything is, that our actions have consequences. We have such a lack of respect for nature and for the environment that we just think, 'Oh things, they will work out in the end.' We don't seem to think about our actions having consequences." Karma and interdependence work in tandem.

And the Dalai Lama, speaking with Greta, also asks us to expand our sense of cause and effect, to think beyond the "very small circles" of ourselves and our own families. Even thinking in terms of countries, he says, is not big enough. "The reality is, individual human beings' life depends on the community. In today's world, the entire seven billion human beings are one human community. So now the time has come, we have to think in terms of all humanity. In ancient times, we lived in small circles, but according to reality now, small circles are unrealistic. So, seven billion human beings depend on each other."

Interdependence means *of course* our actions have consequences, and of course the more of us there are to act on this planet—a number that has grown exponentially in the past decades and centuries and has now passed to eight billion—the more and greater consequences we will have. Interdependence means we're not separate from nature, and we're not separate from each other, and any civilization that runs against the reality of interdependence is delusional and, well, no wonder it can't hold up. Denying and disrespecting our interdependence, no wonder we've created, on the one hand, societies of people who can't get enough, and on the other hand, people who truly don't have enough. No wonder the Arctic is melting, land is flooding, species are dying, and the Earth is burning. Interdependence means *of course* we cannot carry on with business as usual and not expect to be affected by the consequences of our own actions or by the actions of every other jewel in the universe in response.

Vandana Shiva has become a hero of mine since I met her through this conversation. Trained as a physicist, she later shifted to interdisciplinary research in science, technology, and environmental policy; and in 1991 she founded Navdanya, a national movement in India to protect the diversity and integrity of living resources. With palpable vitality and a large bindi on her forehead at her sixth chakra (the seat of hidden wisdom and where her third eye would be), she tells a history of separation in the West that haunts me.

Vandana says that she has "read them all," all of the fathers of modern science including Francis Bacon and René Descartes, and in these readings she saw how "the intellectual architecture of disconnection was laid in that period." She's talking about that time—around the turn of the seventeenth century—when "people who continued to believe they were part of the Earth were defined as witches. Most of them were women." Calling it a genocide, Vandana points out that this was happening at the same time as the genocide of Indigenous people of North America, and she says they're related, since both were about "forced separation."

"Bacon said that we have to torture nature and make her our slave," Vandana reminds us of the violent language he used to describe the separation. Francis Bacon was the lord chancellor of England as well as the inventor of the scientific method, and he oversaw the witch hunts there, apparently. Vandana says René Descartes, for his part, claimed "I'm a thinking *thing*." I can hear the incredulity in her voice. "He can't say 'being,' because being means to be alive," she says, underscoring the difference between *being* and *thing*. Descartes conjured a thinking thing without a body, effectively a person not of this Earth, says Vandana, "because the body connects us to the Earth. Just imagine that," she shakes her head. "Just imagine!" According to Vandana, it followed from Descartes's reasoning that others—other people who were seen as too bodily, and other beings, and nature herself now that she had been othered—didn't think as well or didn't think at all. This way of thinking, Vandana says, "didn't just separate us from the Earth. It separated mind from body and created a very artificial idea of what the mind is, very Cartesian, very mechanical, very militaristic, and also very privileged. It denied intelligence to a living Earth and to every one of her nonhuman organisms, every plant, every microbe, every seed." To some of her human organisms, too. Without blaming Descartes for the witch hunts, I take the point that this way of seeing the world and one another has had profound effects on how we relate.

Vandana understands things very differently—very interdependently. "All the way from the tiny molecule to a cell to organisms to ecosystems and the planet as a whole, there is creativity, intelligence, consciousness pervading. In all our spiritual traditions, we don't just see ourselves as materially connected to the Earth. We get our food from the Earth, we get our breath from the Earth, we get our water from the Earth, but that *consciousness* is the currency of a sacred universe. And we are connected through consciousness. So, the desacralization of the Earth went hand in hand with the desacralization of the human being."

Similarly, from a Buddhist point of view, ignorance of our interdependence is the delusion at the root of suffering—all kinds of suffering, from our daily dissatisfactions to colonialism, slavery, persecution, environmental self-destruction, and war. If my interest in what is "out there" only goes as far as how it serves my needs "over here," I set myself up in a struggle to tame or dominate what is out there, whether it's traffic that's making *me* late or a group of people living on land that I see as a resource for *my* country. Not that these are comparable, but ignorance of interdependence forms the root of suffering at vastly different scales. Such delusion lets us spend our days fearing and fighting for our separate selves (or separate tribes, races, or nations) in petty and catastrophic ways, and it perpetuates itself in vicious cycles—feedback loops—that have put us increasingly at odds with one another and with nature, of which we are at the same time inescapably a part. Buddhists also talk about two other primary loops of human suffering—namely, craving and disliking (the latter often referred to as "aversion")— both of which arise from our ignorance of the way things really are and contribute significantly to the climate crisis.

"Ignorance of Interdependence has not only harmed the natural environment," says the Dalai Lama, "but human society as well. Instead of caring for one another, we place most of our efforts for happiness in pursuing individual material consumption. We have become so engrossed in this pursuit that, without knowing it, we have neglected to foster the most basic human needs of love, kindness, and cooperation. This is very sad."[2]

This *is* very sad. This is suffering. Greta feels it, too. Reflecting on the deep depression of her early adolescence when she stopped eating and stopped talking, she says, "One of the reasons was I couldn't wrap my head around the fact that people didn't seem to care about anything, that everyone just cared about themselves rather than everything that was happening with the world. . . . It made me sad."[3]

Permission to feel sad—I'm grateful for this. It's an honest response, resonant with tenderness and appropriate in this moment. I wonder, if so much of the suffering that we inflict on others and on the Earth, and that we experience ourselves, comes from a misguided sense of separation, maybe it will help to come together and discuss. Specifically, let's talk about these three human feedback loops that Buddhists sometimes call "the three poisons": delusion, craving, and disliking. And let's talk about how they drive climate feedback loops, leading us to the current crisis. I wonder, too, since our human feedback loops have consequences for the planet—given interdependence, of course they do—by the same token are they not opportunities for better consequences, if we intervene to turn these loops around?

THE LOOP OF DELUSION

According to the Global Carbon Atlas, global CO_2 emissions have increased by 65 per cent from 1992 to 2018. Around 50 per cent of all CO_2 emitted since 1751 has been emitted since 1992.[4]

The popular idea of cutting our emissions in half in ten years only gives us a 50 per cent chance of staying below 1.5 degrees and the risk of setting off irreversible chain reactions beyond human control.[5]

Greta repeats facts like these in her speeches over and over because she knows that the path to action starts with awareness, and too many people don't know. Okay, so now we know about climate feedback loops, and even if you didn't before you read the previous chapter, there's a chance that you had a pretty good idea that the climate is changing. You may not read the articles, but it's hard to avoid the headlines. My friend Elissa, the behavioral scientist and tree lover, tells me about a recent survey, the first survey ever of youth around the world—ten countries, she says. The survey asked about their outlook on the world and the climate crisis and how they're feeling about it, and it found that 75 percent of young people, teens and young adults, feel the future is frightening. Seventy-five percent. "The researchers were devastated by their findings," says Elissa, and not only that, but the survey found that 56 percent, more than half of these ten thousand youth, feel that "humanity is doomed."

I'm shocked but not really surprised, since young people have been so visible with Greta in the climate movement in recent years—young people trying to shake the adults awake so we can do something before it's too late. So, what about the adults? How's our outlook these days? Here are some numbers from a study of American beliefs and attitudes by the Yale Program on Climate Change Communication (YPCCC), published in late 2019[6] (2021 data is consistent):

- 72% of Americans think global warming is happening.
- 59% understand that global warming is mostly human caused.
- 66% say they are at least "somewhat worried" about global warming. 30% percent say they are "very worried."
- More than half of Americans say they feel "helpless" (53%) or "disgusted" (50%).

A lot has happened since 2019, the year *Time* magazine called Greta their "Person of the Year," she became a household name, and mainstream media started paying more attention. The YPCCC says some numbers in the 2019 report were the highest yet (e.g., "the highest percentage of Americans since our surveys began are 'extremely' or 'very' sure global warming is happening"), and I would expect these numbers to keep going up. In this very important sense, we can say ignorance of the climate crisis is decreasing.

Yet Greta says in a way we're all climate deniers, "because we're not acting as if it is a crisis."[7] And some climate scientists who contributed to the latest IPCC reports are arguing that no more reports are needed—some are even calling for a climate science strike, since they say we know enough about the climate by now and the focus should be on action.[8] It's a tense time to be alive, caught as we are in this gap between what we know and what most people are doing.

Which I think brings us to delusion in the Buddhist sense of not seeing things as they really are, not seeing our interdependence, not seeing the real causes of suffering. Let us be clear, this conversation is not just for Buddhists, as the Dalai Lama readily admits; it's for all humans. Those of us continuing with our Earth-ravaging business as usual, thinking that the winning and having and "independence" it produces are what make life worth living. Those of us worrying only about ourselves and by extension our friends and family, and what's going to happen to *us*. We in rich countries carry on like we have backup planets, taking resources from other countries and not worrying too much about how it affects them. We keep eating factory-farmed food because it tastes good and—as long as it's government subsidized and we don't count the real costs—it's cheap. Those of us who can afford it plan lives that put us thousands of miles away from loved ones, assuming we can fly to see them as needed, and we go on extravagant vacations far away. Our institutions, laws, and policies for the most part double down on business as usual and the delusion it requires, but the climate attitude surveys, mental health studies, and other indicators don't suggest that business as usual has solved the problem of human suffering. The loop is that ignorance allows us to keep doing business as usual; and the more we do, the less we see things any other way or any alternative to what we're doing.

Just when I'm on the verge of feeling hopeless about all this, Donella Meadows, the systems theorist, helps me see that when ignorance or delusion enables a system to keep going one way, information can be a powerful agent for change. "There was this subdivision of identical houses, the story goes," Donella recounts in an essay about leverage points, "except that for some reason the electric meter in some of the houses was installed in the basement and in others it was installed in the front hall, where the residents could see it constantly, going round faster or slower as they used more or less electricity. With no other change, with identical prices, electricity consumption was 30 percent lower in the houses where the meter was in the front hall."[9] She says "we systems-heads" love this story because it illustrates the power of delivering feedback to a place where it wasn't going before.

She offers another example, from 1986, when the US government instituted the requirement that every factory whose business as usual included polluting the air report their emissions publicly every year. Just from that information—no laws against the emissions or fines to punish the companies for them—emissions dropped by 40 percent in four years and kept dropping. Donella says the new feedback was "not so much because of citizen outrage as because of corporate shame." A chemical company that found itself on a "Top Ten Polluters" list reduced its emissions by 90 percent just to get off that list. Donella also notes that adding (or restoring) information to a system is "usually much easier and cheaper than rebuilding physical infrastructure."

TOP TEN GREENHOUSE GAS POLLUTERS

You've probably heard that a hundred companies are responsible for more than 70 percent of greenhouse gas emissions, or the top twenty companies for 35 percent, or whatever breakdown sounds most shocking. Transparency, as in Donella's "top ten polluters" story, should be a good thing. But interdependency suggests that the takeaway cannot be that it's all the big corporations' fault and it doesn't matter what we as individuals do. Here are the top ten American polluters according to a 2021 report[10] (companies based in the USA, doing damage all over the world):

Vistra Energy
Duke Energy
Southern Company
Berkshire Hathaway
American Electric Power
US Government
Xcel Energy
Energy Capital
NextEra Energy
ExxonMobil

We exist in the same web as these companies, participate in the same systems, as do our banks, food supply chains, governments, and so on. My takeaway is let's not underestimate our responsibility—or our power—as individuals and citizens.

"Human beings," says the Dalai Lama, "are in a sense children of the Earth. And, whereas up until now our common Mother tolerated her children's behavior, she is presently showing us that she has reached the limit of her tolerance."[11] Mother is trying to wake us from our delusion and tell us something. I think the Dalai Lama is inviting us to take this climate crisis as information and let it change us and our systems. Instead of continuing in delusion, we can choose to see things as they really are—the climate crisis, what really makes us happy, and the true interdependence of everything on Earth. And Donella is saying we're not helpless, even in the face of sprawling, complex, and entrenched systems. Knowledge can lead to change.

THE CRAVING LOOP

"When I look inside and see that I am nothing, that's wisdom. When I look outside and see that I am everything, that's love."[12] David Loy, a Buddhist scholar and climate activist, offers this quote from the Indian sage Nisargadatta. David is talking about craving and how it relates to interdependence, specifically to our delusion that we're not interdependent, the false construct that our individual selves are separate. "Because the separate self isn't true, it is fundamentally insecure," he says, "and we experience this as a sense of lack, which we try to fill"—with money or fame or sex or power or potato chips—on the misguided assumption that the thing we lack is "out there." David makes the connection between this individual inner sense of lack and a collective sense of lack, owing to our collective sense of separation from the natural world, a societal delusion. We try to fill this collective sense of lack with economic and technological growth. Individually and collectively, then, we find ourselves caught in feedback loops of craving. "Why," David asks, "is more better, if it will never be enough?"

Speaking of motherly advice, there's the one where your mother asks you to reflect on what the implications would be of what you're doing, or considering doing, if everybody did it. This is Mother effectively sticking up for interdependence, it seems to me, and it might be a helpful way to think about the human feedback loop of craving.

"What if everybody did it?" In this spirit, the Dalai Lama likes the following thought experiment: Add the populations of the two most populous countries (currently almost three billion people combined), pick something people expect to have or do in rich Western countries, like owning and driving a car, and multiply. "Imagine," he'll say, "three billion more cars on the road!" Or even more, as societies become more affluent and families can afford two, three, or more cars.

Imagine three billion more people's water bottles and take-out containers going into landfill. Imagine three billion more people demanding golf courses, recreational space travel, or every latest gadget or fast fashion. Imagine three billion more people eating factory-farmed meat. Clearly Western consumerism cannot be the standard. We will need new, sustainable standards everywhere for climate justice to be realized. It's only fair that we ask ourselves, individually and collectively, whether what we're doing would be sustainable if everyone were doing it, because what Vandana calls "the globalization of greed" is destroying the planet. Something's got to give.

What is that something? Joanna Macy says it's "Industrial Growth Society" that's got to give. Joanna is another systems theorist—and deep ecologist, and Buddhist scholar, and still-active ninety-three-year-old environmental activist as I write this. "The truly operative word is *growth*," she says. "This is a political economy that sets its goals and measures its success in growth. Grow in what? Wisdom? Health? Longevity? Creativity? One thing only. Alas, one thing only." The one thing

Industrial Growth Society cares about is profits, says Joanna, rubbing her thumb and first two fingers together in the sign for cash.

Industrial Growth Society is a runaway feedback loop. "The living systems of Earth are coming apart," she continues, "under the onslaught of the Industrial Growth Society scrambling for the last dollars." We've known for several decades that we've been taking raw materials out of the Earth that cannot be renewed, or faster than the Earth can renew them. "People feel this," Joanna says, "you don't have to be an economist. I mean even a third grader can see that you can't keep growing in a limited world."[13]

This is what Greta is talking about when she talks about "fairy tales" of growth, as she did so pointedly in her UN speech: "People are suffering. People are dying. Entire ecosystems are collapsing. We are in the beginning of a mass extinction. And all you can talk about is money and fairy tales of eternal economic growth."[14] How dare we, indeed. But if Industrial Growth Society is the only society you have known, it's hard to believe things could be different.

When Donella and her colleagues outline an alternative paradigm to Industrial Growth Society point by point, I feel like I can breathe:

Not: Stopping growth will lock the poor in their poverty.
But: It is the avarice and indifference of the rich that lock the poor into poverty. The poor need new attitudes among the rich: then there will be growth specifically geared to serve their needs.
Not: Everyone should be brought up to the material level of the richest countries.
But: There is no possibility of raising material consumption levels for everyone to the levels now enjoyed by the rich. Everyone should have their fundamental material needs sat-

isfied. Material needs beyond this level should be satisfied only if it is possible, for all, within a sustainable ecological footprint.

Not: All growth is good, without question, discrimination, or investigation.

Nor: All growth is bad.

But: What is needed is not growth but development. Insofar as development requires physical expansion, it should be equitable, affordable, and sustainable, with all real costs counted.

Not: Technology will solve all problems.

Nor: Technology does nothing but cause problems.

But: We need to encourage technologies that will reduce the ecological footprint, increase efficiency, enhance resources, improve signals, and end material deprivation.

And: We must approach our problems as human beings and bring more to bear on them than just technology.

Not: The market system will automatically bring us the future we want.

But: We must decide for ourselves what future we want. Then we can use the market system, along with many other organizational devices, to achieve it.

There's more where this came from, but I'll end it for our purposes on this emphatic note of interdependence:

Not: Industry is the cause of all problems or the cure.

Nor: Government is the cause or the cure.

Nor: Environmentalists are the cause or the cure.

Nor: Any other group (economists come to mind) is the cause or the cure.

But: All people and institutions play their role within the large system structure. In a system that is structured for overshoot, all players deliberately or inadvertently contribute to that overshoot. In a system that is structured for sustainability, industries, governments, environmentalists, and most especially economists will play essential roles in contributing to sustainability.[15]

It sounds reasonable, doesn't it? Possibly even within reach? Yet this truth-telling (as Donella and her colleagues call it) has me nodding and shaking my head at the same time—nodding in excitement at the possibilities; shaking because I know the poisons run deep. And there's another poisonous loop that we should talk about.

THE DISLIKING AND FEARING LOOP

"This is a cry for help," Greta says, "to all of you who choose to look the other way every day because you seem more frightened of the changes that can prevent catastrophic climate change than the catastrophic climate change itself."[16] The third poison in Buddhism is usually translated as "aversion," which would be the perfect word except I think it's too stiff, not commonplace or dynamic enough to match the wild, practically constant ride of our aversive thoughts and feelings. *Disliking* would be another word for it. Just think about how much of our experience in a day is like a pendulum swinging back and forth between liking and disliking—or wanting and resenting, craving and hating, loving and fearing . . . Aversion comes in many flavors.

It's easy to get caught in feedback loops of disliking and, as the Buddhist scholar Stephen Batchelor puts it, "fearing that we'll be hurt by what we dislike."[17] Greta is right to call out fear. I'm afraid of what the future will be if we continue down the path of Industrial Growth Society, afraid to think about it as long as I'm sure there is only bad news and doubt that there's anything I can do. So I don't do anything about it or talk to anyone about it, and as a result I don't change my behavior or learn from anyone about how things could be different. I don't hear about success stories or even promising possibilities or get any ideas from anyone about better ways to cope. I don't find a way into any conversation about the climate crisis let alone into any movement for change that makes sense to me, which leaves me alone with my fear. And so it grows. Disliking reality and hiding from fear in "business as usual," even though deep down I know business as usual is taking us to what I'm most afraid of.

At the same time, if I'm being honest (as Greta asks us to be), when I allow for the possibility of a difference that could make a difference, I'm afraid of that, too. Should we all be vegan? What if you really like cheese? Is this the end of road trips or exotic vacations? Of plane travel to see family? Of having or aspiring to have any nice things? Will anything (now I'm spiraling out) ever be carefree or frivolous or hilarious again? I know there are worse fears to have, and it is not lost on me that the people who are lucky enough to have such perks to worry about doing without are the very people doing the most harm; the very people who ought to ask the question: What are the differences that will make a difference? I've been afraid to ask. And so, without new input, this feedback loop, too, continues.

Consider this input, from Donella herself: "Environmentalists have failed perhaps more than any other set of advocates to project vision." She points out that "though it is rarely articulated directly, the most widely shared picture of a sustainable world is one of tight and probably centralized control, low material standard of living, and no fun." She says that whether this is the case because historically some environmental advocates have had a puritan bent or because most of us in the most polluting countries are so thoroughly conditioned by advertising that we "can't imagine a good life that is not based on wild and wasteful consumption," it is, she says, "a failure of vision."[18]

How much energy does it take to have a good life? Researchers at Stanford University asked this question, and after comparing quality of life and energy use across 140 countries, they came back with good news: much less than the average American is using. Americans use nearly four times the amount of energy than what this study found to be the peak for quality-of-life factors such as food supply, infant mortality, life expectancy, prosperity, sanitation, and happiness. The "magic number," according to this study, is 75 gigajoules, where 1 gigajoule is equal to about 8 gallons of gasoline. Above 75 gigajoules is not more—that is, happiness and life satisfaction level off at (or for some of those variables, below) that much energy consumption.[19]

This gives me a vision. Imagine if everyone knew the magic number and we all had real-time gigajoule meters in our "front halls"—more likely on our phones—displaying how much energy we're consuming. It's an example of a future in which nobody's taking anything away from us; rather, we see how things really are and we want them to be different.

In the chapters to come, we'll talk more about disliking, and fear, and other dark feelings that arise with the climate crisis. For now, the Buddha said, it's enough to start seeing our fears more clearly, noticing what feeds into them and what, by contrast, turns these loops the other way, toward courage and a future we can love. There's an antidote, he wanted us to know, for each of these poisons.

A MEDITATION FOR
ECOANXIETY AND CLIMATE DESPAIR

Dekila Chungyalpa

Director of the Loka Initiative at the Center for Healthy
Minds at the University of Wisconsin—Madison, working with with faith leaders around the world to build com-
munity-based solutions for environmental and climate
issues

The first time I led a meditation was at a large environmental
conference in California almost ten years ago, when an audience member asked a question after I gave my talk. When the
mic came to her, she could barely get the words out through her
tears: "How can we bear to sit here and breathe when we know
that the corals around the world are dying right now in this very
moment?" To care about our planet today is to carry suffering, a
manifestation of our longing to protect that which we love best.

Like so many of my peers in that conference room, I was
familiar with the symptoms of what is now called ecoanxiety or
climate distress. However, I was fortunate to have been able to
describe my grief to His Holiness the Karmapa, the head of the
Karma Kagyu lineage of Tibetan Buddhism, and in response,
he had given me a meditation to practice that helped me enor-
mously. Not once up to that point had I imagined that I would
lead meditations for anyone. Growing up in a devout Buddhist
family, the daughter of a Sikkimese nun and teacher, I learned
what kind of qualities and training that role requires. I was and
continue to be absurdly unqualified. And yet that raw and honest

question—*How can we bear to sit here and breathe . . . ?*—and the knowledge that this practice helps drilled through all my deficiencies and doubts and left me with no choice but to offer a meditation practice for the audience gathered there. I have gone on to incorporate meditation as part of the resilience work I do as an environmentalist ever since.

May this practice help us and heal us while we strive to protect this sacred Earth and the entirety of life she has made possible.

Emaho!

In this meditation, we will cultivate our interconnectedness with the Earth through a variation of *tonglen*, the Tibetan Buddhist practice of giving and receiving.

Please start by grounding yourself with the Earth beneath you. Pay attention to how your feet or any other part of your body that is touching the floor is placed. Notice how you are rooted, through a chair or floor, to the Earth and how she literally holds you up—unconditionally, effortlessly, compassionately.

This practice requires that you access the emotions of ecoanxiety or climate distress—grief, anger, vulnerability, sadness, fear—and open yourself up to experiencing them. I ask that you observe how these emotions arise, in what manner and intensity. Pay attention to the shape, color, size, any aspects that give an emotion form. Where does it arise in your body? What are its characteristics? If an emotion overtakes you and washes you away, that is all right. Simply bring yourself back to your purpose of observing, as many times as you need to.

When you have a good hold of the characteristics of your emotions, acknowledge them with respect. Your emotions are a completely valid response to an existential threat to you and your loved ones. It means your inner warning system is working and that is a good thing. So, whatever losses you have witnessed or anticipate, whatever emotions you have suppressed or reacted to, take time to let them all flow out of you and into the Earth. Acknowledge and let go.

Notice your incoming breath—the air entering your nostrils, your mouth, filling up your belly. That oxygen that keeps you alive is coming from forests and oceans, from plants and phytoplankton, from all over the world and from outside your window. Rest in the awareness of this physical manifestation of the Earth's compassion for you.

Every aspect of you right now, the air that fills your lungs, the clothes that you wear, the food you ate today, all of that comes from outside of you. This ever-present, life-encompassing, compassionate Earth sustains you. You are part of this effortless cycle of give-and-take. You are participating in an exchange with the elements, with other living beings, with the Earth herself. With each inhale, breathe in the Earth's compassion and with each exhale, breathe out gratitude.

Relax here in this indivisible connection with all that surrounds you; breathe in compassion, and breathe out gratitude.

Now comes the hard part. Visualize a place or being or community you love that is suffering from climate and environmental harm. It could be a river, a species, the community you belong to, or even the Earth herself rotating in space. Resting in and rooted by the compassion and gratitude you hold, I want you to access your intention, your motivation to alleviate the suffering of your beloved. Now, when you inhale, breathe in their suffering; and when you exhale, breathe out your compassion.

This can sometimes bring up fear, or you may be swept away by grief. If that happens, simply go back to grounding yourself in the Earth's support. When you're ready, come back to inhale the pain and suffering, exhale your compassion and healing.

You can practice this for as long as you feel comfortable and at ease. Do not force yourself; you can always come back to this stage another time.

When you are ready, I would like you to return to the earlier exchange of compassion and gratitude. However, this time you will reverse the direction. Let yourself inhale the Earth's gratitude for your existence; and when you exhale, offer the compassion and love you have for her. You are inextricably connected with her in every moment and there is no division here.

Wonderful. As you emerge from this practice, please set the intention to try it again the next time climate distress or ecoanxiety arises. You can also compress the practice and simply rest in the give-and-take of compassion and gratitude for short moments throughout your day. The work you do is critically important for safeguarding the Earth and all the life she carries. I hope this practice strengthens your inner resilience as you go forward.

THE DAWNING OF THE AGE OF ENOUGH

I hear people say, "People don't change." A friend of a friend said exactly that, moments after he ordered the vegetarian tasting menu at a fancy restaurant that specializes in fish. My friend who was there said this man wasn't vegetarian the last time she'd seen him. The man who said people don't change had changed! We could say, "Easy for him," as he happens to be very wealthy and, no doubt, vegetables go down easier when they come from fancy restaurants and private chefs. Yes, but the fact remains that if you look for examples, you see that people can change. People do change. Science shows that our behavior can change; even our brains and our genes can change. Individuals *and* societies can change.

Roshi Joan Halifax, a Zen priest, anthropologist, and my old friend, offers a historical comparison. Quoting Mariel Nanasi, the executive director of New Energy Economy, a nonprofit organization based in Santa Fe, New Mexico, Roshi Joan reads solemnly,

> We are at a crossroads. We either face the very real possibility of a planet on hospice by an energy system that is the epitome of capitalism on steroids with extreme exploitation and racism at its core, or a profound opportunity to shift at the very basis of our economic system that we haven't seen since the abolition of slavery.[20]

The hospice metaphor is close to Roshi Joan's and my hearts, as we both specialized in end-of-life care. Noting that the first two hundred years of economic life in the United States were based on slavery and the second two hundred on fossil fuels, Roshi Joan says she believes "this next two hundred years must be based on renewables, if we're to survive." Like abolishing slavery, then, this is the right thing to do now; it is, she says, "the moral crisis and imperative of our time."

Others put this moment on a timeline from the Agricultural Revolution ten thousand years ago to the Industrial Revolution three hundred years ago to whatever we would call the revolution that needs to happen now. Joanna Macy says it's already happening, this third revolution of our human journey, a transition from Industrial Growth Society to a life-sustaining society that she calls "the Great Turning."[21] I adore Joanna and revere her work and her writing, but I wonder if we can come up with a more specific—and less melodramatic—name for it? The reality is dramatic enough that I don't think it needs any spin. Others are calling it the "Ecological Revolution," but I suspect that sounds niche to people who don't realize that ecology has something to do with them. Granted, the revolution depends on this realization, but while we're trying to make converts, what if we simply called this next era "the Age of Enough"? As in, enough for everyone, enough but not too much; having enough and feeling like we are enough and enough is enough: the Age of Enough. Doesn't that sound like something we can embrace? Doesn't it speak to everyone who is worried about resources running out and our world ending?

It is fitting in the context of feedback loops, though, to talk about "turning." What will it take to turn the destructive human, climate, and other ecological feedback loops in a life-sustaining direction? Joanna says it happens in three dimensions:

▶ Holding Actions in Defense of Life: Actions to slow down the destruction being wrought by the Industrial Growth Society; political, legislative, and legal action as well as direct action to buy time for systemic changes to take place.
▶ Sustainable Structures: Ways of doing things, new sets of patterns of collective behavior, and ways of construing our relations with each other at every level. (Here she underscores that some of these "new" ways of doing things could and should be very old ways that we recapture.)
▶ Shift in Consciousness: A basic cognitive revolution, a recognition that our planet is a living system, not just a stockpile of resources to be extracted and a sewer for our waste to be pumped into.[22]

"And we belong to it," Joanna says of this planetary living system, "like cells in a living body." She goes on, "I've seen it again and again, as people acting in defense of life wake up to the grandeur of who they really are and find sources of strength, synergies, grace beyond what they could have expected." The revolution is already happening, she says. The question isn't whether it happens but whether it's happening fast enough.

People say that people don't change or that we can't change fast enough. I don't want to take the risk of assuming we're doomed therefore there's nothing we can do. Isn't that just delusion—with the highest possible stakes? Because we don't know that we're doomed no matter what we do. We just don't know. "Despair," notes David Loy, is just another "head trip about the future."[23] The truth is, we don't know; not in the sense that science doesn't know anything, of course, but in the sense that all we know is that we can try to help make things better or they will surely get worse. The choice, as the Dalai Lama says, is whether to be complicit in a self-fulfilling prophecy, an existential feedback loop—giving up, because we're doomed; continuing with business as usual and dooming ourselves—or to really see the possibility of what can be. The choice is the poisons of delusion, craving, and disliking, or their antidotes, which we'll talk about in chapter 4.

We've barely begun to change. Just think what we could do.

PART TWO

Capacity

WE MUST REACT ACCORDING TO THE SITUATION. It is not sufficient to pray to God, Jesus Christ, Buddha, or Allah. We cannot just rely on higher powers. Who created these problems? It was we humans ourselves. And so, it is up to us to solve the problems we have created. The Buddha was very clear about this when he said that we humans are our own masters. Things depend on our own thoughts, our own actions.

—THE DALAI LAMA

RESTORING NATURE IS PERHAPS, it's not only a solution to the climate crisis; it's also a solution to the biodiversity crisis and so on and then we no longer have the possibility to choose between different kinds of actions. We now need to do everything we possibly can and then restoring or rewilding nature is perhaps one of the most important things.

—GRETA THUNBERG

3

EARTH'S CAPACITY

Let the Earth Do What the Earth Does

"To achieve a safe global climate, we must refreeze the Arctic. To refreeze the Arctic, we must cool the Earth. To cool the Earth, we must reduce the heat-trapping gases in the atmosphere." Bill Moomaw, the climate scientist and policy expert, is explaining what it would take to reverse the climate change humans have made. The revelation, to me, is that nature has the power to heal herself if we would only let her, and climate feedback loops hold the keys.

"Current feedbacks will continue and raise global temperatures even if we halted all emissions today," Bill continues with gentle urgency, "because the current amounts of heat-trapping gases in the atmosphere are so high. So, what we must do is simultaneously reduce our emissions and increase the capabilities of natural systems to remove more carbon dioxide than we are releasing." Forests especially, all the scientists in this conversation seem to agree, can turn things around for us if we allow more of them to grow—plant trees, yes, but as we saw in chapter 1, also let old-growth forests be. Recent studies show that forests have the potential to store twice as much carbon as they do today. At the same time, we need to protect existing wetlands and boglands because, as with forests, we need to keep the carbon already stored in them out of the atmosphere and let these living systems do the work of decarbonizing the atmosphere that they naturally do. If we do both these things, if we can sufficiently reduce emissions *and* allow nature's own processes of removal to lower greenhouse gases, the same feedbacks that got us into the climate crisis will get us out of it.

I make a note to remember this when bad news looms large and I'm feeling small. The industrial growth mindset has been so hell-bent for so long on conquering nature and seeing her as the problem, it has forgotten that nature is quite ready to work with us when we take our place in some more natural order. If you're perhaps someone who has lived in a city all your life, someone who likes modern comforts and conveniences, who maybe hates camping; someone who has never thought of yourself—have maybe even scoffed at the idea—as an "environmentalist," scientists and Buddhists alike suggest considering this: you don't have to "return to nature," because you never left. This doesn't mean whatever anyone does is "natural," however, in the sense of harmless; on the contrary, it requires us to consider how we affect nature, since we are inescapably a part of it.

My friend Stephanie Higgs tells me about something a mentor of hers once said to a group of aspiring counseling psychologists. "Many of us were taking classes at the time and learning about different theories of psychotherapy—psychodynamic approaches descended from Freud, theories informed by existentialist philosophy, Maslow's hierarchy of needs," she recalls. "Between taking in these different ways of understanding the human condition and practicing how to apply them, our novice heads were spinning. We were arranging chairs for a group session for support-group facilitators, and our supervisor cut through the noise with a simple statement, saying something that has stayed with me for twenty-five years. She said, 'A space is either therapeutic or it's not.'" Twenty-five years later, hearing the author Paul Hawken speak,[1] Stephanie and I realized he was saying something similar about nature and our place in it. Paul wants us to know, individually and collectively, that our headspaces and living spaces, the systems we build and the ways we go about our days, are therapeutic or they're not, if I may paraphrase. He uses the terms *regenerative* and *degenerative*, hence the title of his recent book, *Regeneration*; but it's the same idea, an invitation to consider the impact we have in the spaces we share, from our homes and workplaces to our cities and social media to the atmosphere and planet. We, in relationship with the rest of nature, have the capacity to harm or heal.

THE CONVERSATIONS in part one of this book were about harm; now I want to talk about healing. I'm aware that the separation of this part's two chapters on capacity into "Earth" and "human" is artificial. I don't mean to further the othering of nature by structuring it this way—I expect Earth's capacity to be full of human activity and human capacity to hinge on better ways of relating to Earth. Nonetheless, this chapter spotlights nature's considerable regenerative powers, for those of us who don't know and are maybe somewhat alienated; while the next chapter singles out humans in the sense the Dalai Lama meant—that as unprecedented troublemakers on this planet with a particular ability for conscious intention, we have a special role to play in regeneration.

Philip Duffy, another climate scientist at Woodwell who is committed to helping inform public policy, says that "ideally, what science can do for us in this conversation is illustrate different possible futures." Let's talk about what the future might look like through examples of the Earth's capacity to heal and our capacity to, as Bill Moomaw simply puts it, "let the Earth do what the Earth does." And as we hear about these local examples, let's keep in mind what we know—from scientists and other wise people—about the interdependence of everything and everyone on the planet: the fact that since greenhouse gases mix through-out the global atmosphere, therapeutic action everywhere in the world decreases warming everywhere in the world. Anywhere we cut emissions, stop deforestation, and regreen the Earth helps to slow, halt, and even reverse climate feedback loops; to lower temperatures, regenerate snow and ice cover and increase reflectivity in the Arctic, refreeze permafrost, strengthen the jet stream, and heal our planet.[2]

REGREENING THE EARTH

"It is extremely important," the Dalai Lama says, "to plant new trees and protect the ones already growing around you."[3] One thing about being the Dalai Lama, I think, is you can bring your weight to a simple statement like that and make it sound freshly momentous. In December 1990 at Sarnath, the place in northern India where the Buddha gave his first teachings and the first Buddhist community was founded, the Dalai Lama gave a speech and handed out fruit tree seeds.

"Since I too have a responsibility in this matter," he said, meaning to protect the environment and "to see that the present and future generations of mankind can make use of refreshing shade and fruits of trees, I bought these seeds of fruit-bearing trees with part of my Nobel Peace Prize money to be distributed now, to people representing different regions (all the continents of the world are represented here)." The Dalai Lama said the seeds were apricot, walnut, papaya, guava, and many other kinds of trees suitable for different geographical conditions, and that they had been blessed. Still, "experts in respective places should be consulted on their planting and care and, thus, you all should see my sincere aspiration is fulfilled."[4]

Where, I wondered when I read this speech, are these trees? If the seed recipients planted them soon after and they survived, they have been growing for more than thirty years. If the Dalai Lama gave you a blessed seed for a fruit tree, wouldn't you track it? Wouldn't you draw it or paint it, if you have the ability, or take photographs or record it in some way? Wouldn't you want to tell its story? Share it on social media (once that came along)? I put out a call but didn't hear back. If you're one of the people who got one of these seeds and you're reading this, will you please tell me? At any rate, thirty-two years later, it's spring as I'm writing this, each chapter so far marked by the trees outside through their progression from bare branches to buds, blossoms, and increasing amounts of green. I'm trying to do them justice.

The climate scientist George Woodwell sums up the solution to the climate crisis when he says we must "make a transition away from fossil fuels and into a new green Earth. And it really does," he emphasizes, in line with the Dalai Lama's promotion of tree planting, "require a green Earth." So, let's talk about regreening.

"I grew up in the countryside," Wangari Maathai says. She is talking about the central highlands of Kenya. "And as a small young girl there was a *huge* tree that was near our homestead, and next to our tree was a stream." Wangari's voice is warm, rich, and deep like good soil on a summer morning, and she has a way of *landing* on important words for emphasis that gives her speech an irresistible rhythm. "My mother told me," Wangari continues, "'Do not collect the firewood from the fig tree by the stream.' I said, 'Why?' And she said, 'Because that tree is a tree of God.'"

"I didn't know what she was talking about, but I would run there and collect water for my mother." Wangari died in 2011, but I'm watching film footage of her activism and old interviews,[5] and I'm spellbound. That voice. "The stream actually came out of the ground, gushing up from the belly of the Earth. Now sometimes there would be thousands upon thousands of frog eggs," she remembers. "They're in black, they are brown, they are white, they're beautiful. I didn't know they were frog eggs. I would just see these beads, and I would put my little hands underneath and try to lift them in the belief that I could put them around my neck and decorate myself. And I would spend hours trying to lift them up. Here I am, and I am so small and I am playing with frog eggs and tadpoles! Between the fig tree and the stream it was beautiful," Wangari pauses. "I guess it was a tree of God."

Wangari says it was not until she went to college that she made the connection between the tree and the stream—the rain caught by the forest canopy's millions of leaves, dripping to the ground, percolating through soil that is held in place, no matter how hard it rains, by a blanket of fallen leaves; soaking down underground where it feeds water reservoirs that follow root systems and push back up through weak places in the earth where they emerge above ground again as streams. When Wangari came back after college, in the 1960s, to the land where she grew up, she says, "I discover now the place of God was in a church. A stone building had been put up. That's where God was. So this tree no longer called for the respect, it no longer inspired awe, it no longer was protected. They had cut it. And sure enough, the stream had also disappeared. And if the stream dies, the frog eggs, the tadpoles, the frogs and everything else that lived in those waters disappears. And we can no longer go there and fetch the water."

Wangari is the first woman in East Africa to have earned a doctoral degree, in biological sciences. In the seventies, teaching at the University of Nairobi, her research took her into the field where she saw deforestation, loss of soil, and the disappearance of streams. Rural women told her they didn't have firewood to cook their food and they didn't have enough water. Recognizing these were symptoms of bigger problems, "That's what gave me an idea, why not plant trees?" The women she was talking to said they would, but they didn't know how. "And that started the whole story of let's learn how to plant trees."

Wangari's whole story has a whole story before it, too. It goes back to colonialism in East Africa, starting in the late nineteenth century, when the British cleared forests to make way for their settlement and agriculture, cash crops like coffee and tea. Later, in the 1950s, they tore down more forests looking for the independence fighters—the Kenyan Land and Freedom Army (or "Mau Mau," as the British called them)—who hid there; and building camps to confine anyone they suspected of aiding or abetting the rebels. Unfortunately, deforestation continued under the new governments after Kenyan independence in 1963. So Wangari founded the Green Belt Movement, in 1977, starting with a few rural women learning to gather seeds from surviving native trees around them, cultivating them in nurseries, and showing one another what they were doing until thousands of women in villages around Kenya were working to reforest their land. Green Belt gave them a small amount of money for every tree that survived, but the women were learning and could see for themselves the many benefits of planting trees.

"The movement started as a tree planting campaign," says Wangari, "that's how we enter into communities. But it is a little more than just the planting of trees." She's being modest, clearly, because she explains that by "a little" she means, "it's planting of ideas, it's giving people a reason why they should stand for their rights, it's giving them reasons why they should protect their environmental rights; and giving them reason why they should protect their women's rights." Green Belt gave civic and environmental seminars around the country to educate and empower people with knowledge about sustainable foodways, nutrition, democracy, self-advocacy and peaceful protest, conflict resolution, and yes, planting trees.

Wangari and her fellow activists were beaten, surveilled, many of them tortured as political prisoners for standing up to Kenya's second president, Daniel arap Moi, whose dictatorial government ruled for twenty-four years. "One of the tactics that the government uses is to make people fear authority," Wangari says. "When the women started, nobody was bothering them because nobody took them seriously. You know, who takes women seriously? Then sometime in the course of the years the government realized that we were organizing women, so they started interfering with our organizing, and they demanded you have to have a license, you cannot meet—they harassed women a lot."

Wangari says the example of "one little woman" (Wangari) successfully stopping one of President Moi's development projects (one that must have been close to his heart since it included a four-story statue of himself) inspired people to believe that the government could change. In 2002, after a year of protests that started when the Moi government planned to log Karura Forest in Nairobi and grew to a nationwide uprising demanding democratic reforms, development of Karura forestland halted and the Kenyan people finally defeated Moi in an election—and sent Wangari to Parliament with 98 percent of the vote. In 2004, she received a Nobel Peace Prize for her work toppling a dictator and teaching people to plant trees. In Kenya, so far, the Green Belt Movement has planted over fifty-one million trees.

Another story of regreening that I love is happening in more than one place in the world, rain-forest places. Local people whose economic circumstances might have or once did drive them to poach lumber from protected forests are instead trained and paid to guard those forests—the trees as well as the animals that coexist in these ecosystems. Francesco Lastrucci, an Italian writer and photographer, tells the Cambodian version beautifully. Owing to the presence of the Khmer Rouge, for decades, in the Cardamom Mountains and the violence and land mines that they brought there, a vast expanse of rain forest was left mostly untouched. But since the conflict abated and the land mines were removed, some people have been plundering the forest's resources while others race to save the forest before it's destroyed.

Wildlife Alliance, Francesco says, "prioritizes round-the-clock law enforcement and collaboration with local authorities, ultimately providing hands-on protection to around three million acres" of rain forest. At the same time, Wildlife Alliance works to create job alternatives for locals through education, reforestation, and wildlife rehabilitation.

From his base camp in the village of Chi Phat, a three-hour bus ride from Phnom Penh, Francesco observes Wildlife Alliance's effects. Whereas recently as the early aughts most people in the area took part in illegal logging and poaching or slash-and-burn farming for a living, now they have transitioned to better options, since better options exist. Farmers have learned more sustainable ways to farm; people have regenerated forestland by rebuilding the soil and planting indigenous tree species—Francesco reports 840,000 trees and counting. The area has become a wildlife tourism destination, employing many locals as they turn homes into guesthouses, for example, and serve tourists as guides. And poachers, with their intimate knowledge of the land, have been recruited and equipped as protective rangers, armed and patrolling the area "on foot, by motorbike, by boat and by air." While the financial attraction of large-scale development and illegal trades is still a threat, according to Francesco, "with an ever-increasing number of locals working alongside the conservationists, saving the forest is no longer a lost cause."

I'm particularly moved by his account of spending time with Soeun, the caretaker of the release station for the area's rehabilitated wildlife, who used to participate in the illegal animal trade. "I went on several walks with Soeun," Francesco says. "A kind and composed man, he introduced me to the animals as if they were members of his family—one by one, and with profound grace and care."[6]

I thank Francesco for his reporting and the *New York Times* for publishing this story. It's easy to lose sight, amid the usual terrifying climate headlines, of a whole world of good news. I want to sing it from my rooftop: climate awareness has a bright side in that it lets us access the stories of millions of people doing good all over the world. You can find them online and via some of the resources in the back of this book.

After realizing that one of the best parts of speaking about the environment is the people he meets after his talks, Paul Hawken and a team of researchers found literally millions of these stories and collected a selection in his book *Blessed Unrest*. "After every speech, a smaller crowd would gather to talk, ask questions, and exchange business cards," Paul says, and when these business cards started to overflow the drawers and shopping bags in which he stored them, a hopeful book about climate change, poverty, deforestation, peace, water, hunger, conservation, and human rights was born. It is a portrait of an emerging movement and a corresponding database of its millions of constituent organizations, "coherent, organic, self-organized congregations," as Paul describes them, "involving tens of millions of people dedicated to change."

"When asked at colleges if I am pessimistic or optimistic about the future," notes Paul, "my answer is always the same: If you look at the science that describes what is happening on earth today and aren't pessimistic, you don't have the correct data." On the other hand, "if you meet the people in this unnamed movement and aren't optimistic, you haven't got a heart." Paul says it's not one or the other. "What I see are ordinary and some not-so-ordinary individuals willing to confront despair, power, and incalculable odds in an attempt to restore some semblance of grace, justice, and beauty to this world." Climate awareness is caring, and caring, Paul says, "is our deepest impulse of life."[7] The same impulse that leads to all the caring we do in a day for our families, friends, pets, gardens, and neighbors can lead and *is leading* many of us to take action to reverse climate change.

I'm happily shocked when he goes so far as to call this moment a cause for celebration. With story after story of humans working on behalf of one another and our beloved Earth, Paul invites us to celebrate rather than thinking of ourselves as cursed.[8]

One more inspiring story of regreening, a future-facing story, that we could talk about is how Arizona's Tucson, America's third fastest-warming city, is planting a million trees. In honor of her late father, who took her and her five siblings into the Sierra Madre mountains to teach them to ride horses and respect nature when they were growing up, Tucson mayor Regina Romero has joined a global network of public, private, and nonprofit leaders pledging to restore a *trillion* trees around the world. In Tucson, she says, they're prioritizing the low-income communities that lack green space and sometimes air conditioning and need trees the most. The University of Arizona assistant research scientist Adriana Zuniga says, "Vegetation is linked to better air, lower temperatures, less stress; studies show less use of antidepressant medication."

Vegetation also provides a resting place for migrating species, Adriana points out. Monarch butterflies, for example, are among the more than ten thousand migratory pollinator species at risk in the Southwest due to deforestation and drought from climate change. "Their habitats have been fragmented so much from urbanization that every spot of vegetation counts," she says. A million more trees in Tucson will provide a much-needed patch of habitat for the butterflies on their long journey between Mexico and Canada. Drought is a problem for trees too, of course, but city officials say that hardy, native species such as desert ironwood and velvet mesquite can not only survive but also help repair the watershed by letting rain get into the ground, unlike pavement and concrete. As Tucson's groundwater is replenished, streams in the area that had dried up by the mid-twentieth century could start to flow again.[9]

Trees heal.

IF EVERYONE ON THE PLANET PLANTED SIX TREES

Diana Beresford-Kroeger

Botanist, medical biochemist, and author who grew up in Ireland and now lives and conducts independent research in Canada

I have developed a course of action for halting climate change. I call it the "global bioplan," a patchwork quilt of human effort to rebuild the natural world that will envelop the entire planet. It is not the ultimate solution to climate change; it is a means of reversing the damage done and of buying us time to find that solution, of stabilizing the climate long enough to address our destructive behaviors in earnest.

The core of the global bioplan is a simple idea. If every person on Earth planted one tree per year for the next six years, we would stop climate change in its tracks. The addition of those wonderful molecular machines—which pull carbon from our atmosphere, fix it in wood, and bubble out oxygen in return—would halt the rise in global temperature and return it to a manageable level. Three hundred million years ago, trees took an environment with a toxic load of carbon and turned it into something that could sustain human life. They can do it again.

What if you don't have the space or means to plant six trees? You take the first step that you're able to take and you have faith.

A personal bioplan can take a form as humble as a pot on the balcony of an urban high-rise. One beneficial plant—a mint, for example—releases aerosols that open up your airways. That plant does the same thing for the birds and other small creatures, and for the people you love and keep close. The true goal of the global bioplan is for every person to create and protect the healthiest environment they can for themselves, their families, the birds, insects, and wildlife. That personal bioplan then gets stitched to their neighbors', expanding outward exponentially.

Of course, we also need to protect all the forest we already have. It's all well and good to each plant a tree a year, but if we're clear-cutting the Amazon and ravaging the boreal at the same time, the positive effects are greatly diminished.

Here, in this protective effort, are even more avenues by which we can fight climate change. On a larger scale, we can band together to take on government and industry; we can keep informed of plans to destroy forests and fight them at every turn. I have been involved in many such efforts. Even when battling against multinationals, international organizations, and governments, we have won.

On a smaller scale, we can take on the role of guardian and steward within our own neighborhoods and towns. If you have a large tree on your street, make sure your local council knows that you value it. Every opportunity to vote is an opportunity to put someone who cares about forests in a position of greater power and authority.

We are just shy of eight billion people on Earth at the time of this writing. With the forest we already have protected, we would have to add approximately forty-eight billion trees—my tally of six each—to absorb enough tonnage of carbon dioxide out of the atmosphere to halt climate change. Forty-eight billion may seem like a number too high to reach, but there's a simple way to get there: just take the first step and keep on going.

Adapted from *To Speak for the Trees: My Life's Journey from Ancient Celtic Wisdom to a Healing Vision of the Forest*[10]

WE DON'T HAVE TO REINVENT THE WHEEL
IF WE REMEMBER THE WHEELS PEOPLE
HAVE ALREADY INVENTED

"A few years ago," Lyla June recalls, "my elder told me, 'Things are going to change very fast, very drastically.'" Lyla June is an Indigenous musician, scholar, and community organizer of Diné (Navajo), Tsétsêhéstâhese (Cheyenne), and European descent, and a great storyteller. Apparently this elder was speaking both prophetically and quite realistically. "One of our water pipes had burst," Lyla clarifies. But beyond pointing out the obvious plumbing problem and its fast, drastic effects—within twenty-four hours, the toilets were backed up and the horses didn't have water—she knew her elder was making a bigger point. "Look how fragile the system is," she heard him saying.

This "long story short" inspired Lyla to enter a PhD program to study Indigenous food systems and land management. From her research of what native people were doing in the Americas prior to Columbus's arrival, she has gathered many stories of people living in very different relationship to Earth than the European colonizers who came much later.

"One of the examples I like to give is the kelp forests of the Heiltsuk Nation in British Columbia, whom I admire quite a bit," Lyla says. The Heiltsuk people still cultivate food this way, she explains, but before they were decimated by disease epidemics in the 1800s, they had even greater kelp forest systems. By hand-planting kelp along the coastlines, the Heiltsuk increase the surface upon which herring fish, "a little silver fish," can lay their eggs. The kelp forests sequester carbon, while the Heiltsuk people eat the herring eggs. For a special delicacy, they anchor hemlock branches in the water with rocks, like upside-down trees, "and when the herring lay their eggs on the hemlock boughs, they give it a special taste." Lyla says she tried the hemlock herring roe a couple of years ago—"I'm from the desert, so it was a bit of an acquired taste"—but orcas, sea lions, salmon, eagles, wolves, and bears love the herring roe as much as the Heiltsuk people, according to Lyla, and "so on and so forth up the food web."

Lyla says the Heiltsuk people think of the herring as relatives, and the kelp forests as a gift, because every year the herring give them the gift of eggs. She contrasts this to the "kill style" of extraction practiced by Europeans and their descendants, whereby commercial fishers "catch thousands of herring, cut them open, drop out the eggs, and throw the bodies back in the ocean"—the fish are "objects, not relatives, to extract the eggs and throw them out."

Lyla has collected many more examples of sustainable cultivation, from a Chesapeake Bay oyster fishery that Algonquin nations kept going for ten thousand years to clam gardens around Quadra Island in the Pacific Northwest. From fish weirs built by Indigenous peoples of Bolivia and rich topsoil in the Amazon that Lyla says didn't happen by accident to a "food forest" grown by Shawnee people three thousand years ago in what is now Kentucky.

What's a food forest? Lyla says that scientists know, from analyzing fossilized pollens in soil cores taken from ponds in Kentucky, that the pollen composition changes in remarkable ways that they can date, because the deeper the sediment the older the pollen. Looking at these cores that are several feet deep, providing about ten thousand years' worth of information, they can see from about three thousand years ago, "all of a sudden," as Lyla tells it, "an influx of chestnut pollen, black walnut pollen, hickory pollen, oak pollen, swamp weed and goosefoot, which are edible species, and even a number of medicinal plant species." Lyla's research doesn't stop at reading scientific articles; she goes into the field—"I had the great pleasure of actually visiting this pond"—to help her picture the fruitful forest that surrounded it, up there on the hill and probably extending down to the bottom lands. Fossilized evidence of the food forest continues for thousands of years, up until the nineteenth century, along with evidence of certain land management practices. Fossilized charcoal shows up with the pollen, for example, indicating regular, low-intensity burns. From these soil samples and from pictures of huge, old-growth chestnut trees before they started disappearing in the twentieth century due to chestnut blight, Lyla says we can tell that the Shawnee peoples managed this food forest all that time.

Lyla says old-growth forests are a hallmark of Indigenous land stewardship. Citing the Menominee forestry project in Wisconsin, the Amah Mutsun oak groves in California, the Shawnee food forests in Kentucky, and her own ancestors in Arizona tending ancient yellow pines, she makes the point that Indigenous peoples "were profound forestry scientists, and in some places we still are."

"We were all pruning the land, clipping the land, burning the land, manicuring the land, getting our hands dirty and being connected to these processes, which in English we call 'food systems,' but really what it is, it's an exchange of life." She continues, "It's giving life, it's taking life, being a part of that transfer of calories and energy." This kind of transfer, remarkably, has the net effect of augmenting biodiversity and storing carbon. She mentions astonishing numbers from a recent United Nations report, that Indigenous people, who are about 5 percent of the Earth's human population, oversee 80 percent of Earth's biodiversity on their lands. So, Lyla concludes, the sustainable revolution depends not only on new renewable energy technology to cut our greenhouse gas emissions but also on recollecting what Indigenous peoples have known for a very long time. It depends on "healing the tissue of the Earth, healing the relationships in the Earth and healing our own understanding of how to be a good relative to everything around us."

"It's not as hard as you might think," Lyla adds.[11] "If you have the right action for a couple of seasons—the Earth actually wants to be full of life. That's its nature. We actually have to work to get it to not do that. So if we just get behind her and breathe life into what she's already doing, it can change pretty fast." But, she says, we need the tools, skills, knowledge, "and most of all the wisdom that supports all of these things."

Another person who recognizes the value of Indigenous knowledge, skills, and wisdom is Kyle Whyte, a professor of environment and sustainability at the University of Michigan and a member of the Potawatomi tribe. "My tribe," part of the Anishinaabe people, Kyle says, "was relocated a long time ago to Oklahoma, but I currently live in the Great Lakes region, which is our homeland. And most of my work has been devoted to empowering Indigenous people as leaders, as scientists, as knowledge holders, as organizers to address climate change." For an example of mutual, interdependent land stewardship, Kyle asks us to "consider wild rice."

"Many Anishinaabe communities live where they do in the Great Lakes Basin because their ancestors were given spiritual guidance to cease migrating from the East when they reached the place where food grows on water." The Anishinaabe word for this food is *manoomin* (wild rice), and Kyle describes a web of responsibilities that grew around manoomin that make it, even today, so much more than a staple of their diet. Responsibilities of older people to teach younger people "to respect, care for, harvest, understand, and prepare manoomin"; responsibilities of younger people to learn these things and thereby participate in family life. Responsibilities for participating in ceremonies that "honor the close connections between manoomin and Anishinaabe society"; responsibilities of committees, like "ricing committees," that are, as Kyle puts it, "accountable to the community for maintaining flourishing subsistence economies."

"Taken together as a web," Kyle says, reminding me of Indra, "all these responsibilities constitute important aspects of the very meanings of family, friendship, trust, character, intergenerational relations, identity, and accountability in many Anishinaabe communities." In turn, manoomin also has responsibilities. "Manoomin is responsible for nourishing humans," Kyle explains. "Manoomin is considered a living being with a spiritual character all to its own. More importantly, manoomin is one of the sources that bring people together into the relationships of family, friendships, trust, and so on. Manoomin motivates these things, which is why Anishinaabe people respect it as a living being with a spiritual character." So the Anishinaabe people steward and guard the manoomin, and the manoomin stewards and guards them, too. And the web of responsibilities, of mutual relationships, expands to include the water that manoomin depends upon to grow, the rivers and rain that feed the lakes, fish like sturgeon that play their own part in the manoomin-human ecosystem, and so on.

At the same time, Kyle reminds us that it's not always sunny for humans, however hard we try to live in balance with the planet. "Earth's beings and systems can engender violence, destruction, and misery." How do we stay in relationship with a reality that doesn't fit our expectations? In a word, we change. Kyle says Anishinaabe communities—facing the climate crisis, racist politics, and industrial pollution—are developing new responsibilities in addition to their traditional ones, in the form of alliances with other groups to protect their relatives (human and nonhuman), political relationships at all levels of society, and "integrating Anishinaabe knowledges with scientific research and teaching scientists to be morally accountable to Anishinaabe communities." Our plant, water, and fish relatives, Kyle points out, "have the power to heal and create new relationships among people."[12]

WHAT CITIES HAVE TO DO WITH IT

Recently a friend of mine in New York City was feeling disillu-sioned with the place she has chosen to live for over twenty years. She told me she thought the immediate trigger was the aftermath of a heavy snowfall the previous weekend, which by the time I was talking to her had deteriorated, the way it does there, from a hushed winter wonderland in the initial hours, like the whole city is sharing a thrilling secret, to lingering dirty piles lining the sidewalks days later, getting grayer and yellower by the hour and strewn with litter and trash while the garbage trucks can't get through the snow-narrowed streets. Add to that pandemic weari-ness, with so many reasons to love the city having been closed or dormant for the better part of two years.

Joining this conversation about climate, my friend was heart-ened by what Jonathan Rose has to say about cities. Jonathan is an unusual real estate developer, investor, and project manager driven by ambitions for what he calls "pervasive altruism" and a "regenerative economy," and he speaks about future cities. Ac-cording to Jonathan, even putting aside visions of what New York City could be, current residents are responsible for only about a quarter of the energy use and greenhouse emissions compared to residents of most cities, thanks to high-density living. "The denser the better," Jonathan says—driving very little and using a relatively excellent mass transit system. ("Even if it doesn't always feel excellent when you're using it," said my friend.)

We talk about how environmentalism traditionally repre-sented itself with images of unspoiled nature and animals in the wild, like the Sierra Club desk calendar my friend got every year when she was a teenager—week after week of scenic photo-graphs without a single human being. Spoken or unspoken, the assumption in modern Western culture, like Vandana Shiva was saying, is that people live outside of and in opposition to nature.

But Jonathan thinks about cities differently. At its best, city living provides an example of humans' capacity to hurt the planet less by pooling resources, putting up with one another on the subway, and sharing space, and he is saying we can build on this.

"We have an economic system that draws from the common good," says Jonathan. "It actually draws from both the ecological common good and from the social common good. It consumes those for profit." No argument here. "So, what would be a system in which profit was created not by degenerating the commons but by regenerating the commons?" Jonathan asks. Imagine if Vladimir Putin had spent the last twenty years obsessing over that question instead of Ukraine.

Jonathan says we'll never achieve a regenerative economy "as long as our system is organized around self-maximization or business maximization, deal maximization, fund maximization, country maximization." He shares a vision of pervasive altruism, a whole new economic system based on compassion, where the success of deals and developments are measured in terms of ecological sustainability and an all-inclusive common good. He wants to add "people" and "planet" to "profit" on the bottom line.[13]

"What are the levers that we can understand that trigger pervasive or collective altruism," Jonathan asks, and answers his own question. His speech tumbles out of him like he's got a lot to say and perhaps not enough time to say it. "Okay, so it begins with a vision. You will never get somewhere if you do not know where you're going." He has googled "visions of future cities" and is sharing the results. "You see a lot of interesting common elements," he says. "The buildings are actually taller and denser. They're hooked up by transit systems that are quiet, green, electric, shared. And they're pervaded by nature. We see this in images from all over the world, in all cultures, that people are yearning to live with nature. They want cities that are more dense, more walkable, more equitable, greener, and full of nature."

Jonathan presents as an affable businessman—he's even wearing a blazer—making his radical proposals all the more surprising. "We can have visions for where we want to go. We can then plan established community health indicators. And then use the tools of government—regulation, investment, incentives, and leadership—to actually control our outcomes, measure the outcomes, compare the community health indicators, and create virtuous circles that co-evolve toward our vision." He says once you have a vision and you turn that vision into a set of measurable qualities—here he shows us the Santa Monica Sustainability Index, by way of example, with processes and qualities like "solid waste diverted from landfill," "traffic congestion," "community involvement," and, of course, "greenhouse gas emissions"—you can continually monitor and take corrective action and continually evolve. It all sounds so reasonable and so doable.

Beyond doable, it's being done. Oslo, the capital of Norway, for a leading example, is showing the world what carbon budgeting looks like. To follow through on the city's ambitious commitments to reduce greenhouse gas emissions (a 95 percent reduction from 2009 levels by 2030), every department of the municipal government is accountable; every policy or plan weighed with its environmental impact in mind. The journalist Nick Romeo describes active construction sites so quiet with all-electric machinery that people enjoy outdoor dining while heavy equipment tears up asphalt nearby. In Oslo, all municipal construction projects will be zero-emission by 2025. "Look at Oslo," Nick says, "and you can begin to see what life will look like in a city that's serious about its obligations to the future. The shifts are subtle but pervasive, affecting everything from cemeteries, parking, and waste management to zoning, public transportation, and school lunch. Rather than waiting for a single miraculous solution, Oslo's approach encourages a dispersed, positive shift."[14]

To balance its carbon budget, the city offsets expenses like new electric vehicles and equipment with income from eco-friendly incentives like increased tolls and parking fees for non-electric cars. (The Norwegian Environmental Agency having assessed that 33 percent of the city's emissions come from private vehicles.) As residents of Oslo get used to incentives like these, Nick reports, they gradually escalate "from nudges to prohibitions." Grants to encourage people to convert from oil-fired furnaces gave way to an outright ban, and the newest budget requires the city to become an emission-free zone, electric vehicles only. To illustrate the power of such strict policies, Nick gives the hypothetical example of a delivery company that might have been willing to eat the cost of expensive parking for its gas-powered vans, whereas a ban could motivate the company to go electric and replace its fleet.

We should not underestimate the power of cities. Many changes in Oslo have been so politically popular that the federal government is starting to cooperate and adopt carbon budgeting for decisions that affect the whole country. Along with Barcelona, Berlin, Los Angeles, Milan, Montreal, Mumbai, Paris, Rio de Janeiro, South Africa's Tshwane, and Stockholm, Oslo is one of eleven cities from the C40 participating in a carbon budgeting pilot program; the C40 being a network of cities around the world that together represent 20 percent of global gross domestic product (GDP) and a thirteenth of the human population. The world's biggest cities have higher populations and economic outputs than some countries. "Often small and nimble enough to avoid the gridlock of national politics," Nick says, yet "also big enough to make a meaningful difference," cities can hit the sweet spot of climate action. I've heard the ambitiously green mayor of Boston, Michelle Wu, make the same point. When people see what's possible in these cities, hopefully national governments like Norway's will be inspired. And voters can demand it.

Speaking of Barcelona, I'm excited to learn about "superblocks" from Anupam Nanda, a professor of urban economics and real estate in the United Kingdom. First introduced to the city in 2016, superblocks are areas of nine square blocks—three by three—with no access by cars. Traffic has to drive around them. Superblocks, Anupam says, reduce pollution from vehicles where people live and "give residents much-needed relief from noise pollution."[15] Within a superblock, pedestrians and cyclists can roam freely. Anupam mentions a study carried out by the Barcelona Institute for Global Health predicting that if the more than five hundred superblocks Barcelona has planned are realized, they would encourage people to use public transportation to the tune of 230,000 fewer private vehicle journeys per week. Air quality would be significantly improved on the car-free streets, and temperatures would be cooler as the space formerly taken by cars is turned over to green spaces and trees. Designed to entice people outdoors, superblocks also offer the physical and mental health benefits of a more active lifestyle.

Anupam concedes that superblocks as designed for Barcelona may not be replicable, exactly, in every other city due to "a great deal of variation in scale, population density, urban shape and form, development patterns, and institutional frameworks." Large cities in the developing world that are "heavily congested with uncontrolled, unregulated developments and weak regulatory frameworks" may not be superblock-ready. But Anupam says that with some adjustments "the basic principles of superblocks—that value pedestrians, cyclists, and high-quality public spaces over motor vehicles—can be applied in any city."

I keep thinking about Tucson's million more trees. It's both a massive effort and a simple action, simple like the Dalai Lama saying, "It is very important to plant new trees." There are countless examples of cities trying to be better, mitigating and adapting to climate change, but the example of Tucson especially moves me because, along with all the specific benefits of greener cities, it simply reminds me that cities and the people who live in them are as much a part of the Earth as anywhere and anyone, and as such they have everything to do with Earth's capacity. Like a dandelion growing through a crack in a sidewalk, a million-tree planting in an American city is an even bigger living metaphor of our cities' rootedness to our planet. May cities celebrate that connection and give us fewer reasons to feel like we need to escape.

TECHNOLOGY ALONE WON'T SAVE US

"We live in a strange world," says Greta, "where we think we can buy or build our way out of a crisis that has been created by buying and building things."[16] In certain circles, for example, people are talking about building trillions of mirrors in space, ostensibly to reverse the effects of global warming by literally reversing enough of the sun's rays—space mirrors to replace the Earth's natural ice and snow mirrors as we watch them melt. From listening to the scientists and the Dalai Lama, however, I'm with Greta. I'm wary of far-fetched, astronomically expensive technological "solutions" that don't get to the roots of our problems and, critically, don't yet exist. Can't we talk instead about Earth's capacity to heal and cool and how we can let her? Wouldn't it be easier and more certain to remember what we already know and do what we already know how to do? We know that the past seventy years of industrial development and intensive agriculture have badly degraded natural systems. As nature is showing us that we can't go on producing the food we need by continuing to wage war on her, interest is growing in decarbonization through regenerative agriculture, agroforestry, organic farming, and rewilding.[17] For those of us in the richest and most wasteful countries, wouldn't it be easier to simply do—to buy, build, drive, etcetera—less? Is it necessary, advisable, or even ethical to invent space mirrors or other Frankensteinian geotech monsters even if we could, when nature already knows how to capture carbon and we already know how to regenerate soils, wetlands, peat bogs, mangroves, and seagrasses and so on—to plant trees and leave old-growth forests alone?

To the geotechnology bros, says Vandana Shiva, "the problem is the sun." She shakes her head. "This disconnected mind takes nature and says, 'You're the problem. You're the problem and I have to discipline you even more.' Poor guys," she says, "I feel sad for them." Sad for Silicon Valley? Vandana explains, "I feel sad for them in the midst of this awakening of humanity to a living Earth, a living consciousness. Even at this time, they're trying to stamp out consciousness, stamp out interconnectedness, stamp out life. And I think it's a bit of a desperate measure." Instead of all this stamping and inventing, Vandana says we could return to "our sacred place." She elaborates, "If physics has woken us to the quantum coherent universal interconnectedness, and ecology has woken us to the quantum coherence of biological and ecological interconnectedness, is it not time to put the petty games of petty minds aside?" Like "kids who never grew up," says Vandana, for some, "the Earth is a LEGO set." Her advice is inviting: "Grow up and be part of the sacred universe. It's a joy."

The Dalai Lama and many others in this conversation avidly follow creative developments like biological building materials and technologies that turn seawater into water for drinking and irrigating crops. Industrial Growth Society has done so much damage already that net-zero emissions and regreening, some scientists say, will not be enough; nature may need our help recapturing and storing carbon, and this nascent technology deserves major investments to scale up.[18] But His Holiness offers the following caveat:

This does not mean that I believe that we can rely on technology to overcome all our problems. Nor do I believe we can afford to continue destructive practices in anticipation of technical fixes being developed. Besides, the environment does not need fixing. It is our behavior in relation to it that needs to change. I question whether, in the case of such a massive looming disaster as that caused by the greenhouse effect, a fix could ever exist, even in theory. And supposing it could, we have to ask whether it would ever be feasible to apply it on the scale that would be required. What of the expense and what of the cost in terms of our natural resources? I suspect that these would be prohibitively high. There is also the fact that in many other fields—such as in the humanitarian relief of hunger—there are already insufficient funds to cover the work that could be undertaken. Therefore, even if one were to argue that the necessary funds could be raised, morally speaking this would be almost impossible to justify given such deficiencies. *It would not be right to deploy huge sums simply in order to enable the industrialized nations to continue their harmful practices while people in other places cannot even feed themselves.*[19]

Emphasis mine. The Dalai Lama wants science and technology that is wise, motivated by "the full richness and simple wholesomeness of our basic human values";[20] science and technology not poisoned by our delusions, cravings, and dislikings but in service of our capacities for facing reality, caring for others, and feeling like we are enough and have enough. This is why he founded Mind & Life.

4

HUMAN CAPACITY

The Necessity of a Sense of Efficacy

I'M SCROLLING THROUGH INTERNET lists of species we've lost to environmental destruction and climate change. The lists—and the extinctions—are a thing:

> "23 Species from 19 States Lost to Extinction" (Center for Biological Diversity)
>
> "Protected Too Late: U.S. Officials Report More Than 20 Extinctions" (*New York Times*)
>
> "11 Recently Extinct Animals" (treehugger.com)

And so on. Most of these lists come with full-color photographs, images so close-up and in such high resolution that I can forget for a moment that the animals aren't alive in front of me, let alone extinct, and I find myself trying to look these strange, amazing, and incredibly varied creatures in the eyes. Well, hello there, giant tortoise. And how are you, gray-blue bird, who looks like you're about to say something? Oh right. I remember what I'm doing and why I'm doing it; this photo shows a bird no longer found in the wild. As I mentally zoom out from the close-ups, life appears less lively without these fellow species plodding and soaring with me, and I feel deeply sad. It's too late for the Western black rhino, the Yangtze River dolphin, the splendid poison frog. They're gone.

But Greta has something to say that speaks to my deepest feelings about the climate crisis: it is not too late for us humans to "wake up and change."[1] She is emphatic. We have more than enough facts about the climate crisis, painstakingly gathered and reported by climate scientists over the past forty-odd years. More recently, we also have the facts right in front of our eyes, ears, and noses as we live with the smell of nearby wildfires, the sight of flooded neighborhoods, and the sights and sounds of a movement—millions of people around the world marching and sounding alarm bells with their voices and handmade signs: *Wake up and change!*

We have the capacity to wake up. And we have the capacity to change.

It's easy to take this the wrong way, like a finger wagging, making me feel bad about flying to see my family or not being vegan all the time. *Wake up and change.* It can backfire, if it only reminds us that we're not sure people *can* change—or not enough people, not fast enough anyway. And it can seem like a bummer, like the other day when people at work were excited about the first spring shipment of cut flowers arriving at Trader Joe's, and I didn't want to be the one to say, "Hey, wake up and don't have flowers, because they're probably grown with chemicals and overheated conditions that are unsafe for the migrant workers and the Earth, and they're shipped by fossil-fuel-powered vehicles hundreds or thousands of miles."

The *shoulds* pile up while the sense of *can't* threatens to overwhelm us, as we go round and round in our human feedback loops, living like we're the center of the universe, wanting more and more and avoiding the truth about the climate and our own behavior when it's bad news.

In chapter 2 we talked about our capacity for delusion, craving, and disliking (and fear). If you struggle with any or all of the above or have understandably tried to avoid the struggle by keeping your eyes closed; if you try not to think or talk about climate, or look at the extinct or suffering animals, or click on the scary headlines; if you continue with business as usual because the crisis feels hopeless and, anyway, you don't know what to do, I can promise you two things. First, in this chapter we'll talk about turning around the noxious (and obnoxious) human feedback loops in our lives and society with the age-old remedies of knowledge, cooperation, and compassion, and we'll consider the implications of doing so for climate. Second, we'll allow ourselves to hope, and we'll talk about why hope, even knowing what we know, isn't crazy or naïve. In fact, it is necessary.

I come back to Lyla June, someone who makes me want to wake up and who helps me think about—and believe in—our capacity to change.

Lyla is telling a story about her Indigenous ancestors. I am rapt, as I often am when Lyla speaks. She appears to be glowing, and I don't think it's just that soft-focus Zoom thing. Her story takes us to Chaco Canyon in present-day New Mexico, but she is talking about a civilization that existed there from about 900 to 1150 C.E. The place where these people lived is famous—a National Historical Park and UNESCO World Heritage Site—because of their extraordinary architecture, enough of which remains to continue impressing visitors to this day, nearly nine hundred years later.

"What most people don't know," according to Lyla, "is that we made many mistakes at this place." I notice how she counts herself among her ancestors, and I'm touched by the humility as well as her sense of interconnection.

She tells us about the caste system that developed in this civilization, with the priests at the top and all the people much lower down whose labor built all those incredible buildings. "We also had a lot of playing God," she adds, "doing things that humans aren't meant to do," like unsustainable water management and bringing people back from the dead. Society went on like this for some 250 years until, according to traditional archaeological accounts, all these people mysteriously just left.

Lyla says it's not a mystery to her people. "Nobody stops to ask us, but we actually know what happened to us," she states matter-of-factly. "Creator sent us a drought, and this drought gave us the courage to change when nothing else would." The young people in particular, she tells us, said "enough." Now the story is sounding awfully, wonderfully familiar, right?

"If it wasn't for that drought," she explains, "we would have continued our ways of dehumanizing each other. We would have continued playing God, depleting and manipulating water and land." The drought, as she tells it, was a supreme opportunity: "The drought was our profound gift." Even more, she says it wasn't a fluke, or a one-off, or a myth. Lyla points out that human history has produced many stories of world-destroying droughts, fires, and floods, and she believes they're true, as true as the opportunities they have given us to evolve rather than perish—to wake up and change.

"What I'm trying to say," she says, "is that the climate crisis, as we know it, is tragic. There's no two ways about it. It *is* tragic. We've been given every canary in the coal mine possible, every red flag, every wake-up call, every warning, and we haven't listened. We have had chances to choose humility over fear and change, and we haven't taken those chances. The way I see it, this is our drought. This is it."

She talks about a word in the Diné language, *k'é*—pronounced with a clicky *K* and a breathy *EH*, it sounds musical to my ears. *K'é* means interdependence, kinship, love, and being there for each other. She says her people are grateful for that drought nine hundred years ago because it prompted them to dismantle the caste system and rebuild their society to be equitable and sustainable, a society based on *k'é*. She concedes it might be controversial to say so, but to Lyla our climate crisis could likewise be a catalyst, a crucible—a gift.

"We are going through a hard passage in the journey of humanity and planet Earth," Joanna Macy agrees. Lyla's story about what happened in Chaco Canyon is an example of what Joanna means by a Great Turning. "This is a mammoth adventure," says Joanna, her eyes so wide and such pellucid blue they could be a metaphor for waking up and seeing clearly. "But in every myth, legend, adventure story that our culture has treasured and passed on, the hero or heroine is called at some point to look straight into the face of the monster or the medusa and not turn to stone."

She goes on, her voice rising and calling us like a preacher's: "Through our suffering with our world, we are seeing the immensity of the life in us. Saint Francis, for example, the saint of ecology in Christianity, he didn't run from the bad news of his time. He walked with Lady Poverty. He kissed the leper. He kept his eyes open to the devastation of war and was a prisoner of war. And by *being* that pain, collectively with his brothers and sisters he could sing the "Canticle of the Sun" and open to the blessedness and holiness of life and get a sense of *una immensa vita*."[2] Amen, sister.

I heard a similar sentiment from Edward Maibach, the director of the Center for Climate Change Communication at George Mason University. I first met Ed when he gave a talk at Mind & Life's Summer Research Institute, our annual gathering of great scientific and spiritual minds. That was 2021, the first time the Summer Research Institute focused on the climate crisis. "I've been doing this for a long time, and for thirteen, fourteen years now I've focused exclusively on climate change, because I've come to see climate change as the greatest public health threat that we face," Ed says, leaning forward. "But I've also come to see climate change as potentially the greatest public health opportunity that we face." Threat and opportunity, emergency and possibility; the crisis has two faces like Janus, the Roman god of endings and beginnings.

What kind of opportunity, exactly? He explains, "Those of us in public health, we understand how important economic stability is; the attributes of our neighborhood; our physical environment; whether or not we have access to good education, good food, a stable community context, and good health care. But a stable climate is fundamental to all those things. So, I really have come to see a stable climate as the most fundamental determinant of health." In other words, addressing climate change will have far-reaching, positive impacts on so many things we already care about. Ed's research and practice focus on activating trusted groups of professionals like health-care providers, TV weathercasters and other journalists, and educators—people in positions to, in grant application language, "enhance public and policymaker engagement in climate change." Which, I think, is another way of saying our conversations with doctors, nurses, teachers, and media figures have the capacity to help us wake up and change.

That humans have the capacity in any given moment to wake up and change happens to be a very Buddhist idea. We can sleepwalk through life, reacting to what comes at us according to old habits and unexamined assumptions, daydreaming of the past or fantasizing about the future; or we can wake up, see things as they are, and make choices about what to do next. As we talked about in chapter 2, Buddhist texts and teachers suggest, as does Greta in her inimitable way, that things as they are, are interconnected. We are interdependent, and not just with some people and some things but with everything and every being. Buddhists are encouraged not to take this as scripture but to wake up and look at the evidence. And when we see how things really are, including how we affect the systems in which we live—which also exist in the endless web of causes, effects, and interdependence—we might see reasons to change. We might even find ourselves acting different already. Change can come surprisingly naturally when we know the reasons for it in our bones. This capacity to get real is our superpower against ignorance, and it works on the big scale of the science of climate feedback loops as well as in our day to day.

I recently woke up to an example of interdependence. I was picking up litter along a road near my house in rural Virginia. It's a new Sunday ritual for me, going out with a couple of empty bags and returning when they're full of the bottles, cans, fast-food wrappers, and other random stuff that people throw out of car windows. I'm realizing that it does more than make the neighborhood look a little nicer. Standing with someone's empty plastic bottle in my hand, it occurred to me that it's not just an object, and I started thinking about everything that went into getting it here, from the car driving on this road and the person who bought it and drank the water and threw the bottle out the window, to the store it came from, the stocker who put it on the shelf, the cashier who rang it up, and the truck that delivered it to the store, back to the bottling plant that bottled the water, filling this and millions and millions of plastic bottles like it, that are made and thrown away every day without a thought. I imagine the chemicals that went into making the plastic that made the bottle and how those phthalates and BPA are getting into our bloodstreams and those of the animals and insects exposed to the plastic. I recall that it takes up to a thousand years for plastic to decompose, yet the chemical fragments of plastics continually leach out and seep into our soil and natural water. I am reminded of being in India where water in plastic bottles is the primary potable water available and seeing and breathing thick, toxic smoke simmering throughout villages as they burn the plastic bottles to get rid of them. I imagine the fossil fuels used to power the bottling plant as well as the trucks and cars that moved this bottle. I wonder where the water in the bottle came from—a river or lake, or did it come from a city water source and just got dressed up with a more appealing label? Were there filters or chemicals used to "clean" the water? Then I think about all the human lives that were involved in all these processes, from the engineers, plant workers, truckers, storekeepers, and their families to the people running the companies and getting rich simply from making plastic and bottling water. The fact that I would put this plastic bottle in my recycling bin when I got home didn't seem like a drop in the Earth's bucket compared to all that, and for the first time I felt like I was seeing a single-use container clearly.

All our actions, big and small, take place in webs of interdependence like this, and as Greta so persistently says, we all have the capacity to wake up—and show up.

SPEAK THE TRUTH

Greta Thunberg

LAST WEEK WELL OVER 4 MILLION PEOPLE in over 70 countries striked for the climate. We marched for a living planet and a safe future for everyone. We spoke the science and demanded that the people in power would listen to and act on the science. But our political leaders didn't listen. This week world leaders gathered in New York for the UN Climate Action Summit. They disappointed us once again with empty words and insufficient action. We told them to unite behind the science. But they didn't listen. So today we are millions around the world striking and marching again. And we will keep on doing it until they listen. If the people in power won't take responsibility, then we will. It shouldn't be up to us, but somebody needs to do it.

They say we shouldn't worry, that we should look forward to a bright future. But they forget that if they would have done their job, we wouldn't need to worry. If they had started in time, then this crisis would not be the crisis it is today. And we promise: once they start to do their job and take responsibility, we will stop worrying and go back to school, go back to work. And once again, we are not communicating our opinions or any political views. The climate and ecological crisis is beyond party politics. We are communicating the current best available science.

To some people—particularly those who in many ways have created this crisis—that science is far too uncomfortable to address. But we who will have to live with the consequences—and indeed those who are living with the climate and ecological crisis already—don't have a choice. To stay below 1.5°C—and give us a chance to avoid the risk of setting off irreversible chain reactions beyond human control—we must speak the truth and tell it like it is.

In the IPCC's SR15 report that came out last year it says on page 108, chapter 2, that to have a 67 per cent chance of staying below a 1.5°C global temperature rise—the best odds given by the IPCC—the world had 420 gigatonnes of CO_2 left to emit back on January 1, 2018.

Today that figure is already down to less than 350 gigatonnes. With today's emissions levels, that remaining CO_2 budget will be entirely gone within less than 8.5 years.

And please note that these calculations do not include already locked-in warming hidden by toxic air pollution, nonlinear tipping points, most feedback loops, or the aspect of equity, climate justice.

They are also relying on my generation sucking hundreds of billions of tonnes of CO_2 out of the air with technologies that barely exist.

And not once, not one single time, have I heard any politician, journalist or business leader even mention these numbers.

They say let children be children. We agree, let us be children. Do your part, communicate these kinds of numbers instead of leaving that responsibility to us. Then we can go back to "being children."

MONTREAL, SEPTEMBER 27, 2019[3]

OUR CAPACITY TO COOPERATE, SHARE, AND NEED LESS

Monks and nuns are living examples of our capacity to coexist in close community and to need fewer material things. I happen to know a few personally, but the humble and secluded nature of their lives also means that they aren't highly visible or familiar examples to laypeople living in secular society. From the point of view of rapaciously consumerist and hyperindustrial society, the monks and nuns seem to be the ones missing out on modern creature comforts like, say, tasting menus with wine pairings or video games.

But as Greta and the most famous monk on Earth agree and as they have each said to the world repeatedly, humans in general and especially in rich countries need "a whole new way of thinking."[4] Because, from another perspective that emerges from this chapter's conversation, it's not the monks and nuns in their simple robes who are missing out, or Greta in a sailboat because she doesn't fly, it's us, or those of us caught up in feedback loops of wanting, and shopping, and envying, and striving, and competing. Mistaking conveniences and status symbols for needs, we don't realize our capacity to share, cooperate, and need less—or different—than we thought we needed. To start to turn our craving around, it might help to speak directly with some living examples of generosity and simplicity.

"Taking a breakfast from my balcony of my little hut, that is three meters by three, and that's all I need. Even if a fairy came to give me three material wishes, I have nothing that I need," my friend and Mind & Life colleague Matthieu Ricard says in his jovial French accent, and I believe him. Matthieu's unofficial title is "happiest man in the world," and he has the smile lines around his eyes to show for it. "Why should we need all those stuff? Why should we always keep on wanting more and more and more?" Why indeed. "This is a crisis of the superfluous. And also of the disparities, the inequalities that have been increasing the last thirty years in all the developed countries. The climate crisis really boils down to altruism versus selfishness. We don't have three or five planets, we just have one, so we must do better with less."

I think about this. Just as we have plenty of evidence from physical scientists to understand how we're affecting our planet, social scientists—just like all the wisdom traditions for thousands of years before them—have been trying to tell us that wanting and having things, beyond a certain point, doesn't make people happy. They even tell us what that point is, and it's strikingly modest by rich-country standards. So why? Or rather, why not find a whole new way of thinking?

Matthieu is a Buddhist monk at Shechen Monastery in Kathmandu, Nepal. If three meters by three sounds cramped to you, you should see where it is—and you can, on his website. For Matthieu is also a photographer, and he wants to show us a photograph of the view from the balcony of his "little hut," above the clouds in the Himalayas, looking across a valley that looks softly lit from within by morning light to layers of mountains made ethereal by the mist. So, depending on how he defines home, depending on his way of thinking, it is not small at all.

"There's a Tibetan saying that 'When you have one, if you

want two, is to open the door to the demon,'" Matthieu offers, opening a door to what would be, for so many of us in the overdeveloped world, a whole new way of thinking. "Tune to a beautiful landscape, or the face of an innocent child, or the face of a wise teacher, now you want to bring the best of yourself. And these things are also the most fulfilling."

We know this, but we forget. Hearing Matthieu say it makes me want to start a most-fulfilling list of my own and refer to it often, to defend against the everyday bombardment of societal pressure and advertising that tries to tell me and sell me what it wants me to believe is most fulfilling. A list to help me separate the needs from the wants and the goodness from the nonsense: A funny and loving text from a dear friend. Dipping in a cool mountain brook after a long, hot hike. When my dog snuggles against my leg. Gardening, putting my hands in the soil, weeding, watching the seeds grow, then eating the fresh vegetables that I've grown. Voting—and standing outside the polling site with a sign that says, "Vote for Love." Offering someone a ride as I'm driving by when a downpour starts and they don't have an umbrella, that big feeling of helping someone with something big to them, even when it's not a big thing to me. A funny and loving text *to* a dear friend.

Matthieu concludes, "Selfishness, me, me, me all day long makes you miserable, makes everyone miserable. It's a lose-lose situation, while altruism is a win-win."

The Dalai Lama makes a similar point, illustrated by another vivid picture, of materialism infecting the Tibetan way of thinking as the Tibetan community, exiled to India, settled in. As he describes it, "Some insensitive Tibetans in Tibet wear outfits adorned with tiger, leopard, and otter skins. It appears that guardian deities of Tibet decorated with such outfits are influencing them. They imitate the clothes worn by these deities, without even having an iota of knowledge about them. Such behavior, indeed, makes every one of us feel embarrassed."

Yikes. I personally would like to avoid making the Dalai Lama embarrassed for me. He continues, "Many Tibetans are also fond of flaunting their wealth by wearing heavy gold rings on their fingers. Such rings studded with precious stones are very popular in Tibet. Some even wear such enormous rings that it is difficult for them to move their fingers. Their fingers appear as if they are injured and plastered with bandages. Mother Nature has gifted us fingers in such a way that we can move them freely. So it is better to keep them as they are."

See what His Holiness did there? These guys in the tiger furs were thinking big rings were what they needed, but the Dalai Lama flipped that way of thinking. Who needs jewelry? Suddenly our naturally nimble, bare human fingers are the things to have, and we already have them. Suddenly they are enough.

"These days in India . . ." He is not finished. "Tibetans are not considered to be humble." He explains that in the years since they were exiled from Tibet in 1959, some of them have gone astray and that those engaging in "illegal activities such as murder, smuggling, and trading in animal skins have brought disgrace to the whole Tibetan community. Our community has made progress in the field of education. We have made improvements in our economic conditions also. But instead of being contented with our lives and becoming better human beings, we seem to be becoming worse. If this trend continues, then imagine what sort of a future we will create for our people!"[5]

This reminds me of something Greta talks about often, climate justice, which she says is "absolutely necessary to make the Paris Agreement work on a global scale." She sums it up precisely: "That means that rich countries need to get down to zero emissions, within six to twelve years, so that people in poorer countries can heighten their standard of living by building some of the infrastructure that we have already built. Such as roads, hospitals, electricity, schools and clean drinking water. Because how can we expect countries like India or Nigeria to care about the climate crisis if we, who already have everything . . ."[6]

We who have everything haven't been willing to wake up and change; to share, cooperate, and stop thinking we need things we don't need, like big rings, big cars, new gadgets, redundant shoes, you name it—the latest thing.

"We need a whole new way of thinking," Greta says. Our political and economic systems built on competition, cheating, and winning "must come to an end, we must stop competing with each other, we need to cooperate and work together and to share the resources of the planet in a fair way. We need to start living within the planetary boundaries, focus on equity and take a few steps back for the sake of all living species."[7]

As this sinks in, I understand that the Dalai Lama and Greta are getting at the same idea, from opposite ends. At first, many Tibetan refugees in India really didn't have enough; their material needs were real. But as their economic conditions improved, they needed to know when to stop wanting more stuff. For her part, Greta comes from twenty-first-century Sweden, where most people have so much more than they need. Taken together I think Greta and His Holiness are suggesting that we meet somewhere in the middle, in a place of being contented with our lives—realizing we have enough (because in this place I'm imagining, we all do) and are enough—yet becoming better human beings; a place that is, not coincidentally, sustainable.

Together, as living examples, I think Matthieu, Greta, and the Dalai Lama are saying we have the capacity to find our way to this place but only if we help each other get there.

MY GRANDSONS are four and newborn as I'm writing this. I think a lot about what to tell them about the future and what they will say to me as they continue to develop the language to express themselves and to describe our world. I want them to know that stuff will not make them truly happy. I want them to love our planet and their future as much as I love them, but like so many parents and grandparents these days, I worry with a very heavy heart. So much is uncertain.

Joanna Macy recently reminded me—I'm paraphrasing—that uncertainty is not knowing, not knowing opens the space for possibility, and everything is possible when we don't know. It helps to think about that. I want my grandsons and all the young people in my family and around the world to have that sense of possibility and, at the same time, know that, like unadorned fingers, they are enough as they are, in this moment—beautiful, whole, enough, right now.

THE CAPACITY TO CARE ABOUT MORE THAN OURSELVES

My personal experience, and I know I'm not alone, is that it takes considerable courage to look the climate crisis in the eye. People dislike the climate crisis, to put it mildly, and people are afraid. I am afraid. People who lived through the Cold War are once again living with a kind of existential dread that they had hoped was in the past. It takes courage to see clearly these days, knowing what we now know and feeling how we now feel about the future. It takes courage not to be paralyzed by climate anxiety, to live with climate-related suffering in the world and not ignore it; courage to keep thinking, talking, protesting, and voting—and keep believing that it matters that we do. It takes courage to ask, "What can I do?" not as a helpless rhetorical question. It takes courage to keep trying to imagine a future we can love and do something about it.

Where do we find this courage? My friend Thupten Jinpa, the Dalai Lama's longtime translator, says a key insight of His Holiness is that fear is often rooted in our selfish concerns, as when we fear being judged, disliked, rejected, or otherwise hurt.

Remembering our shared humanity, on the other hand, and choosing a standpoint of compassion leaves little room for fear or alienation. As Jinpa puts it, compassion "instantly opens up space for courage, not to mention a kind of deep-down relaxedness that comes from the lack of self-agenda." Deep-down relaxedness—I like the sound of that.

Jinpa was a Tibetan Buddhist monk himself until the age of thirty-seven when he left the monastery with plans to have a family. He recalls fearing at the time what the Dalai Lama, among others, would think of his decision. I'm not surprised to hear that His Holiness was kind, understanding, and accommodating when it came to it—he is the Dalai Lama, after all. But the general point Jinpa is making is well taken, that shifting his perspective from what would people think of him to how his decision would affect others and what he could do to minimize hurt and reassure them gave him courage.

We fear what's going to happen to *me*; how climate change is going to affect me and my family, my property value, my air quality, my comfort. When we consider how our quality of life depends on access to limited resources, it can stir up fears of *them* versus *us*, too. Again, it helps to think instead about others. As the Dalai Lama said to Greta and the rest of us, "Other animals, you see their daily life: eat, sleep, sex; but we are not so simple. We have much desire, worries, and concerns, urges and cravings and feelings. And too much sense of 'we' and 'they.' I think among the different species of mammals on this planet, we human beings create a lot of good things, but at the same time, we create a lot of problems.

"So now the question is why the human brain, though wonderful, gets so stuck in narrow thinking—firstly, about ourselves as individuals; secondly, of our own family; and finally of our own nation, our own country. This thinking is a very small circle. The reality is, individual human beings' best interests depend on the community. Entire seven billion human beings are one human community, you see, so now the time has come, we have to think in terms of all humanity. Individuals' best interests depend on humanity."

What happens when we change our thinking in this way? Jinpa has written a whole book about compassion, and he says one thing that happens is we get brave. (The book is called *A Fearless Heart: How the Courage to Be Compassionate Can Transform Our Lives.*[8]) The Dalai Lama observes, "Compassion is by nature peaceful and gentle, but it is also very powerful."[9]

I think about Greta's courage to raise her voice to the United Nations assembly, global leaders, and the entire world of grownups who maintain the status quo. "How dare you," she said. I marvel at how, when she started her school strike for climate with the power of a giant compassion, she was a fifteen-year-old girl concerned with the well-being of no fewer others than every living being in the world. Greta admonishes politicians for not speaking honestly about the climate crisis because they're afraid of being unpopular. "But I don't care about being popular," she says. "I care about climate justice and the living planet."[10] And though she began her *skolstrejk* (school strike) alone, by the time she spoke to the United Nations, it felt like most of the world had her back. I hope it felt like that to her. I hope that she feels supported by the courage of everyone who joins this conversation, just as we have felt encouraged by her.

The word in Buddhism for a person who lives with Greta's and the Dalai Lama's kind of other-orientation and courage is *bodhisattva*. The term comes from the Sanskrit for enlightenment (*bodhi*) and sentient being (*sattva*), but in Tibetan "courage" is built right into the word, because when Tibetan Buddhists translated *sattva*, they gave it the extra meaning of "courageous" being.

Jinpa tells us that traditional texts speak of two main aspects of what a bodhisattva does—namely, developing themselves and working for others. Self-development how, exactly? "Self-development partly happens through the practice of Six Perfections," Jinpa says, "including morality, diligence, concentration, wisdom, generosity, and patience." Self-development also comes through working for others, specifically "helping others with their immediate material needs, communicating with others in a pleasant way, sharing insights on how to live in a virtuous way, and embodying such teachings by way of personal example." Our celebrity-obsessed society puts modern-day examples of bodhisattvas like Greta Thunberg and the Dalai Lama on a pedestal, and I can see how if we're perfectionistic about it, we could fail to recognize what these two heroes have to do with us. They're so good at what they do that their lives may seem out of reach. But Jinpa says bodhisattvas aren't superhuman; that is not the point. The point is that we all have the capacity for compassion. However unenlightened we may feel in a given moment, we can remember that each of us has bodhisattva potential, an inner bodhisattva if you will, and we can tap into—or wake up to—this. In Buddhism, it's our buddha nature; in Judaism and Christianity, Jinpa says (he has a PhD in religious studies), it's our inner divinity or inner spark; in Islam, it's spirit, the aspects of the human psyche that reflect the aspects of the Creator.

Taking inspiration from Greta and the Dalai Lama's examples, finding courage in my climate compassion to join this conversation, though it is scary and hard, my inner spark burns a little brighter and I feel excited that I, in turn, in some way, could be an example for someone else.

The climate scientist and communications specialist Katharine Hayhoe points to our capacity for compassion by appealing directly to it. The climate, she says, affects everyone, "no matter who we are and where we live. But it disproportionately affects those who are poor, those who are marginalized, those who are disabled, those who are already living on the edge."

I have a sense of this, but she makes it vivid: "Imagine there's a heat wave that is bigger and stronger and longer and more intense than it would be otherwise," Katharine explains. "Who is most affected by that heat wave? People who have low-income jobs who have to work outdoors, people who cannot afford to pay their air-conditioning bill, people who don't have well-insulated homes or might have broken windows who can't protect themselves from the heat, people who live in unsafe neighborhoods so they don't feel like they can open their window at night to get a little bit of relief from the cool air. These are the people who are most affected." It's true within the United States, she reminds us, as well as around the world.

"People who lack access to basic health care, people who live in extreme poverty below one or two dollars a day, when their crops fail, when their wells dry up, where do they find food for their family?" Katharine asks. "Where do they find water? Climate change affects us all, but it affects the poor and the marginalized more than anyone else. And that is not fair."

Again and again, Greta calls for climate justice: for rich countries, rich people, and corporations with the luxury of choice to do more and pay more while poorer countries and communities raise their standard of living and build some of the infrastructure that comfortable people already have.

Katharine reminds us that the flip side is also true: "Climate solutions benefit us all, just as climate risks harm us all. But climate solutions benefit the poor and the marginalized the most. And that is fair."

She gives us another example: "A lot of rich countries are rich because they have massive fossil-fuel resources that they use to industrialize. Most low-income countries do not have large amounts of fossil-fuel resources. And the few that do, like Nigeria and Venezuela, those resources are typically extracted by large multinational corporations, and the revenue is used to enrich the rich at the expense of the poor. But many low-income and tropical countries have a lot of sun and a lot of wind. And so in 2020—and remember, electricity is highly correlated with human well-being, more so than energy use in general—of the new electricity installed around the world, more than 90 percent was clean energy. Why? Because in many parts of the world, solar and wind energy is cheaper."

Katharine mentions a program called Solar Sister in sub-Saharan Africa that empowers women to become entrepreneurs, selling solar lanterns and solar cells that people can use to light their homes. She mentions places where people are turning human waste into fuel, for example an organization in India that has built nearly two hundred plants that use waste from public toilets to create renewable natural gas or biogas that replaces fossil fuels. She is animated about agricultural solutions, too: "Putting carbon back into the soil where we want it instead of in the atmosphere where we don't, through conservation agriculture, agroforestry, water conservation, through empowering and educating women and girls, especially in low-income countries, through restoring and preserving ecosystems so they support rural economies."

It's easy for some people to be dismissive of compassion when we talk about the climate crisis. It sounds soft. It sounds like a feeling, and the climate doesn't care about our feelings. But this conversation has helped me see that compassion is critical to coping with our fear and is fundamental to a whole new way of thinking required to turn things around. Maybe it's easier to dismiss compassion than to practice it. Let's take that as a challenge.

In 2022, the year of the IPCC's *sixth* Assessment Report on climate change over thirty-four years and the year after the *twenty-sixth* United Nations Conference on Climate Change (COP), we have the knowledge. The question is what do we do with that knowledge? Do we have the capacity to change? How do we turn that knowledge into "right action," as Buddhists put it? For one thing, perhaps the most important thing, compassion. More than a feeling, compassion is a call to action because it taps into a sense of responsibility and requires us to think beyond the immediate and beyond ourselves. It takes courage to open our eyes and unplug our ears to hear this call, let alone answer it. But at the same time, having compassion for one another and all the beings that share this planet will make us brave. As Jinpa says, compassion takes courage, but it also *makes* courage, because having compassion for others frees us from fearing for ourselves. Plus, with this orientation toward others comes the realization that we have eight billion teammates. As the Dalai Lama says, "One humanity."

As Greta says, "Imagine what we could do together if we wanted to. Every single person counts. Just like every single emission counts. Every single kilo. Everything counts."[11]

THE CAPACITY TO HOPE

"I know you are desperate for hope and solutions," says Greta—she feels us—"but the biggest source of hope and the easiest solution is right in front of you, and it has been all along. And it is us people, and the fact that we don't know. We humans are not stupid. We are not ruining the biosphere and future living conditions for all species because we are evil. We are simply not aware. But once we understand, once we realize the situation, then we act, we change. Humans are very adaptable."[12]

Greta's right, humans are very adaptable. So why can't I shake the feeling that I'm being naïve when I talk about hoping for a future we can love instead of fear? The news and the experts pile on reasons to be afraid, to feel depressed, to think we're doomed, so much that it's hard to remember a time when it was controversial to worry about the climate at all, though it wasn't long ago. Now we're at a point where hope seems to have become the more controversial thing. So, let's talk about hope. Through the essayist Rebecca Solnit, I've asked Virginia Woolf to join us.

"The future is dark, which is the best thing the future can be, I think," Woolf wrote in her journal on January 18, 1915, when she was almost thirty-three. I'm thinking, what could she possibly mean?

Reflecting on these lines, Rebecca notes that at the time, "the First World War was beginning to turn into catastrophic slaughter on an unprecedented scale that would continue for years." This wouldn't seem to answer my question, but then Rebecca does.[13]

"To me, the grounds for hope are simply that we don't know what will happen next and that the unlikely and the unimaginable transpire quite regularly." There it is! She continues, "And that the unofficial history of the world shows that dedicated individuals and popular movements can shape history and have, though how and when we might win and how long it takes is not predictable."

Woo-hoo! Worried that I could be cheering prematurely, I don't say that out loud. It might be too soon in the conversation to declare hope the winner, but I feel secretly, if cautiously, exhilarated. Go Rebecca! Go Virginia Woolf.

"Despair is a form of certainty," she says, "certainty that the future will be a lot like the present or will decline from it." And she points out that optimism, as she defines it, has less in common with hope than despair, for like despair, optimism is a false sense of confidence about what will happen. Both optimism and despair, she warns, "are grounds for not acting."

I'm reminded again of what Joanna Macy says about uncertainty, how it is the best ground for the seeds of possibility to germinate, sprout, and grow.

Do people whose life's work is to study the causes and effects of climate change every day feel hopeful? That would seem to be the ultimate test of hope, no? I met Bonnie Waltch, a writer and producer of documentaries, through the climate feedback loop films that Greta and the Dalai Lama helped launch, which Bonnie coproduced. Astonishingly, to me, she assures us that all the scientists they interviewed for those films were (cautiously) hopeful. She recalls the Woodwell Arctic expert Jennifer Francis saying, even as she was talking about how we call it the "New Arctic" now, that we can always make things better than they would be otherwise.

"If they didn't have hope," Bonnie supposes, "I don't think they'd be able to do their jobs, and it's because they care about the planet and humanity that they do the work they do. That really gives me hope, too."

Bonnie herself knows the threat of despair. "When I was researching the topic and discovering all the different stories that we could tell, it terrified me, honestly," she admits. "It prompted me to put solar panels on my house, and research buying an electric car, riding my bike to work." Hearing this gives me mixed feelings of guilt and admiration, as I can't say I have always been able to channel terror so constructively. But Bonnie believes that demystifying climate change to see our part in it lets us see what needs to be done and how we have control. I'm moved by her example.

Take it from Don Perovich, the professor of engineering at Dartmouth College whose research in sea ice geophysics takes him to polar landscapes asking questions about how sunlight interacts with ice and snow. "As a working scientist," he says, "I'm well aware of what's going on in the Arctic. I mean, the defining characteristic of the Arctic is the perennial presence of ice on the ocean, the sea ice. On land, there's glaciers and ice sheets. In the land, there's permafrost. Sea ice is melting, ice sheets are melting, and permafrost is thawing. If we look at my field of interest, sea ice, when I started over forty years ago, even at the end of summer, when it was at its minimum extent, it was still the size of the United States."

Forty years later, he says, "It's as though the entire country east of the Mississippi melted." My first reaction is, *Shit.* But I'm also curious: How does somebody who works so close to the crisis, who's seen all the data, keep doing his work and not lose hope?

"I'm well aware of the problem, but so are other people," Don continues, seemingly having read my mind. "We made this problem. We can work together to solve this problem. There's more and more effort toward solar power, toward electric cars, and other efforts like that. As a scientist, I kind of gravitate toward the technology. But even more important is I'm an optimist because I believe in people." Echoes of Greta. "What we've seen over the past ten years or so is just a total sea change"—so to speak—"in public awareness of what the problem is and how we can all work together to solve it."

Since Don's version of optimism doesn't seem to be founded in false confidence or causing him to be complacent, I gather that optimism to him is what Rebecca Solnit meant by hope. The Dalai Lama, when he's speaking in English, also uses "optimism" to describe a positive orientation to the future that is both active and reality based. "We must be determined and must have an optimistic outlook," he says, "then even if we fail, we will have no regrets. On the other hand, lack of determination and effort will cause double regret. Firstly because the objectives were not realized, and secondly because you feel guilty and regret not having made full effort in the realization of the objectives."[14]

Christiana Figueres was the executive secretary of the United Nations Framework Convention on Climate Change (UNFCCC), and her strong, positive presence and bridge-building skills were instrumental to her success in brokering the Paris Agreement on climate change in 2016. Christiana is one of my heroes. She credits "stubborn optimism" for her perseverance through that process, and I ask her what that means.

"Doing the impossible," she says, which sounds awfully paradoxical to me. She explains, "I inherited the failure of Copenhagen and was told, 'Well, go look in the garbage can to see where that political process is and then see what you can do about it.' And very early on, there was a deep realization that branding something or condemning something as impossible would only be a self-perpetuating reality and that the first thing that had to change here was the mindset." Stubborn optimism is starting to sound more practical than paradoxical—would she care to say more?

"First, let me start with the optimism piece and let me say what we don't mean by optimism. We don't mean a naïve assumption that everything's going to be fine, I don't have to do anything about it. That's totally irresponsible." So, what does it mean? "Optimism is a deliberate choice that we make every single day to face whatever is in front of us, be it in our daily life, our relationships with our loved ones, or the global planetary crisis of climate change. Optimism is an input, not a result of having achieved something. That's not optimism, that's a celebration. And frankly, we don't celebrate enough, so we should celebrate more. But optimism, to me, is not the result of success or the result of an achievement; it's the input with which we go into a challenge." Optimism as an input—I love that.

As for the stubborn part, it's pretty simple, "Stubborn because, predictably, it's going to be tough." Christiana brought together national and subnational governments, corporations, activists, financial institutions, communities of faith, think tanks, and technology providers and got them to agree on a multinational plan for global cooperation to address the climate crisis. So yeah, that's going to be tough. And something worth celebrating, for sure, when it was accomplished. But of course the work isn't done, and it's going to be tough. So, we need to stay stubbornly stubborn *and* optimistic.

OPTIMISM. HOPE. I prefer fewer syllables, other things being equal, and I like that "hope" is a verb. But whatever word works for you, we can agree the distinction is important: hoping and working together for a future we can love is definitely not the same thing as counting on it while doing nothing about it.

Katharine Hayhoe says she is asked almost every single day what gives her hope. "I've turned that around and I've asked hundreds of people the same question here in North America and Europe and beyond. I got a lot of different answers, but they all boil down to people seeing other people act, recognizing that we are not alone, recognizing that I as an individual can make a difference through using my voice, through advocating for change in my place of worship, in the place where I work, in my community, in my neighborhood, in my school, in my social organization, as part of my city or state or country." An evangelical Christian, Katharine is famous for finding ways to engage in dialogue about the climate crisis with people who initially don't want to talk about it at all. Overflowing with clear, practical advice and examples, she is one of the most articulate and compelling climate activists that I have ever heard. The trick, she says, is to talk about what people already care about; that's our "in," because no matter what it is, there's a climate angle. Some people's capacity to wake up and change to cleaner energy, for example, starts with saving money, when they see that it will reduce their monthly utilities bill. Meeting people where they are and being willing to work from there strikes me as hope *in action*. Our motivations can be pretty prosaic, yet Katharine wants us to know that at least for a start, that's okay. Whatever it takes.

Lyla June finds hope for the future in learning from the past. She reminds us that the Great Plains of the United States, for example, were created by people in cooperation with nature, not by nature alone. In some Indigenous lunar calendars, she says, "the September moon is called the Grass Burning Moon, because it was time to burn the plains." She explains, burning recycles the nutrients of the expired plant tissues by turning it into ash. The ash goes into the soil, bringing potassium, nitrogen, and phosphorus along with it. Burning also creates charcoal that, as Lyla puts it, "creates little apartment rooms for microbiomes." So, this kind of controlled burning returns nutrients to the soil, stimulates microbial activity in the soil, supports certain species that only flower after fire, and significantly reduces the chance of uncontrolled wildfires.

If they hadn't burned those prairies, Lyla says, "if you leave the grasslands alone, they will close into shrubs and trees. Grasslands don't live without humans. So, what you have in the wake of these fires is that nutrient-dense grasslands come up, and who loves nutrient-dense grasses? Buffalo. A lot of people think that we followed the Buffalo, but there's increasing evidence that the Buffalo followed us and the prairies we made." Lyla finds something profoundly hopeful in this: "Human beings can be a keystone species." A keystone species, she says, "is one that if you were to remove it from the system, the system would fall apart. An example is a beaver makes a dam, which makes a pond, and all these other species enjoy the water. They create an entire habitat," with their swimmerly little bodies, unstoppable teeth, and trademark tails.

"Humans," Lyla says simply, "are like that. We are meant to create habitat for other species." When she was at the Parliament of the World's Religions a few years ago in Toronto, "there was a Yoruba elder there and he got on stage. And he said in the Yoruba language, the word for human being means 'chosen one,' because we were chosen by the Creator to steward the Earth. We were the species that was chosen to take care of Mother Earth because all the other species needed help. And maybe these big brains and these opposable thumbs are given to us so that we can give to the Earth in a way that no other species can. And case study after case study that I read proves this, that Indigenous peoples the world over were taking up that role. They were taking up that mantle and they were saying, 'Yes, I will be that steward. I will be that warrior. I will be the one who tends the land so well that I augment life everywhere I go.'"

"We've done it before," Lyla says, "and we can do it again."

We can be a species on which the whole interdependent system of life on Earth depends, a species that other species can't live without, instead of a species they can't live with and a species that destroys itself. It's not a new way of thinking, apparently, but it's different. Acknowledging and talking about our capacity to see clearly and think differently—to care about more than ourselves, to share and cooperate more and consume and compete less, to realize that we have enough and make sure that everyone does, to know that we are enough, and to hope and love more than we hate and fear—this is the waking up part that is so critical to change.

If we're worried about sounding naïve, we can put it like this, in Greta's tempered words: "We are failing but we have not yet failed."[15] It seems to me that we need to know this now, while it is still true.

PART THREE

Will

So now the question is why the human brain, though wonderful, gets so stuck in narrow thinking, firstly, about ourselves as individuals, secondly, of our own family, and finally of our own nation, our own country. This thinking is a very small circle. The reality is, individual human beings' best interests depend on the community. Entire seven billion human beings are one human community, you see, so now the time has come, we have to think in terms of all humanity. Individuals' best interests depend on humanity. Happy humanity, healthy world.

—THE DALAI LAMA

I haven't created a movement, I haven't mobilized people, it is me together with other millions of people of all ages, but especially young people. It is we, together, who have done that.

—GRETA THUNBERG

5

HEARTBREAK

The Darkness and the Light

"SOMETIMES HAPPINESS is on the other side of giving up the thing we think we can't live without," my friend Stephanie Tade, who is also my literary agent, observed in one of our early conversations about this book. She has noticed that wisdom and strategies for substance problems have much to tell us about how people change, and she has lived the truth of this. I ask her to tell me more.

"It was a long time ago, when cell phones weren't in the palm of every hand," she begins, "but if some magical being had followed me around with their camera on, and then we watched the video of my life, there would have been little doubt about what was going on with me. The chaos, the hangovers, the recklessness, the drunken driving. But to me, living it, my life just looked like . . . my life. Choices seemed to make themselves. I couldn't imagine things being any other way. The idea of living without beer, or wine, or whatever was unimaginable. Who would even want that? How would I do—anything?"

She recalls how, at the time, she thought alcohol made her life bigger and more fun, but looking back she can see it was the size of an Amtrak bathroom compared to what she has today. "My relationships were a mess; I could just barely hold on to a job. I couldn't be counted on to show up, for anything."

"So, what happened?" I ask.

"So much has been said about 'bottoming out' that it has become a cliché. But really," says Stephanie, "it's a very personal moment, one that has limitless potential for change. Bottoming out is what happens when we allow our heart to break, and it creates a gap from which compassion, insight, and creativity can spring. You don't know what will trigger it. For me, after years of alcoholic drinking and the degradation and the mishaps that go along with that, it wasn't jail, or a gutter, or bankruptcy; it was one Saturday when I slept through the time that the pool was open at the YMCA. It was the disappointment in that, or maybe the better word is *dismay*, because swimming was the one thing I did that made my life seem okay. And that caused a little crack, wide enough that the reality of my alcoholism was able to slip past all my defenses and justifications and habits."

So much, says Stephanie, of what we do comes down to habits. But dismay, desperation, sorrow, grief, shame, or outrage—heartbreak in all its forms—can stop a habit in its tracks and open the way to something different, even if only for a moment. And there can be enough room in that moment for something new, something true and real, to come in. There can be what Stephanie calls "teachability" in heartbreak; it can be a moment that changes everything. In the moment after Stephanie realized she'd woken up too late to go swimming, she remembers the first thought that entered her mind: "I need to make a phone call." And she did.

I tell her that I am glad she did. I would hug her, but we're on Zoom, she in Bucks County, Pennsylvania, and I in Virginia. So we sit together with these feelings for a bit, looking at each other through our screens, then agree that this potential that we're talking about, this possibility in the shock of seeing the prognosis of what is to come if we do not change, could have implications for the climate crisis. We decide that part three of this book will be about the heartbreak of this moment on this planet—a bottoming out of our species, if you will—and how the light gets in.

GRETA SAYS, TALKING TO THE DALAI LAMA, "We are desperate for a raise in spreading of awareness, and we need to tell people about what's happening right now." I've heard her call for more awareness many times, but this time the word *desperate* stands out to me. Whereas phrases like "educating ourselves," "reading up" on a subject, "facing the facts," and so forth point to wrestling with the climate crisis in our heads, desperation is a feeling. This matters, because teachability in bottoming-out depends at least as much on the heart as the head, as the word *heartbreak* suggests. In Joanna Macy's words, "The heart that breaks creates a space big enough for God to move in."[1] Sometimes the crack is wide indeed, and it's going to have to be if it is to hold our grief for life as we know it.

When the British writer and environmental activist George Monbiot saw the movie *Don't Look Up*, a meteor-heading-toward-Earth allegory of people's ostrichlike attitudes about climate change, it reminded him of the time he broke down on a popular morning show in real life, like the actress Jennifer Lawrence's scientist character in the movie. "It was soon after the COP26 climate conference in Glasgow," recalls George, "where we had seen the least serious of all governments (the UK was hosting the talks) failing to rise to the most serious of all issues. I tried, for the thousandth time, to explain what we are facing and suddenly couldn't hold it in any longer. I burst into tears on live TV."[2] George says he was mortified when it happened and still feels embarrassed. "The response on social media, like the response to the scientist in the film, was vituperative and vicious. I was faking. I was hysterical. I was mentally ill. But knowing where we are and what we face, seeing the indifference of those who wield power, seeing how our existential crisis has been marginalised in favor of trivia and frivolity, I now realize that there would be something wrong with me if I hadn't lost it."

Joanna Macy agrees. "You are suffering with your world? That's the most natural thing in the world," she says, "the most wholesome, healthy thing in the world, too."[3] She remembers riding a subway across the Charles River from Boston to Cambridge in 1976, when it first really sunk in. She had attended an all-day symposium on threats to the biosphere. Sitting there on the T looking out at the sailboats in setting sunlight, she remembers thinking, "Yes, we can do it now, we can destroy our world. Hadn't that knowledge been lurking all along in every sight of streaming smokestacks and clearcut forests? That knowledge came out of hiding now, and I had no idea how to live with it."[4]

I want to talk about how to live with this knowledge and these feelings. Joanna tells another story, from a couple of years later, when she was asked to lead a weeklong working group at a meeting of the Society for Human Values in Higher Education, as part of a conference at Notre Dame.[5] The first morning, thirty or so participants gathered for introductions. "Because I could not bear to hear people identify themselves by their rank and academic credentials," she recalls, "I broke with convention and said, 'Please introduce yourself by sharing an image or an experience of a moment when you felt the planetary crisis impinge on your own life.'" What happened after that was "some kind of magic." People spoke briefly but plainly of their pain at what they saw happening to the world, their fears for their children, their discouragement. "It came to me like a thunderclap," she says, the realization that "Oh my God, the suffering can be spoken!" And it wasn't just her. Speaking their suffering brought the whole group together, "unleashing creative energy and mutual caring. . . . Sessions went overtime, laced with hilarity and punctuated with plans for future projects." We may be heartbroken and scared, but in daring to talk about it, we realize we're not alone.

Sharing my broken heart and caring about yours, says Joanna, "delivers us to a new kind of humanity."

The former-climate-journalist-turned-Buddhist-teacher Catherine Ingram describes her own experience coming to terms with the truth of what we're doing—what we've already done—to our planet. The author of an article she titled "Facing Extinction," she is, as that title suggests, unflinching. "As I began to realize the gravity of our situation, I quickly recognized that my own death was not much of an issue."[6] Though she doesn't pretend to have no fear of death, she has already lived a long time and has gone a long way to accepting her mortality. "No," she says, "the despair came from the thoughts about my young great-nieces and great-nephew with whom I am close. All nine of them were under the age of ten when I began to realize that they are not likely to have long lives." Catherine's assessment of the research is very dark, and her outlook could be summed up as *sooner than you think*. "The anxiety and despair into which I sank was such that I became very ill," she recalls. "I developed a massive case of shingles covering large areas of my torso, front and back in two zones (apparently it is rare to have shingles in more than one zone) and I ended up in the hospital." Shingles, she says, is a stress-related illness. "My anxiety and despair had made me physically sick. Once home and bedridden for the best part of a month, I had a chance to consider how unaffordable my fear and anxiety would be going forward. I had to find a perspective that would allow me to access at least some quiet underneath the profound sadness, some whisper that says, 'This is the suchness of things. Everything passes.'"

She acknowledges that, by choice or necessity, people relate to the climate crisis in vastly different ways—for example, hundreds of millions of climate refugees around the world, parents of young children among them, struggling to survive climate change–related catastrophes that have already happened. For them, says Catherine, "any fretting about the future would seem the greatest of luxuries and privileges." They have been forced to come to terms with it. On the other extreme, she knows that many parents currently in more comfortable circumstances cannot or will not talk about the present and future suffering and destruction of climate change. She has had to accept "that virtually no one in my family and few of my friends are either ready to hear this information now or will be prepared to face what is ahead in time."

George Monbiot chimes in. "As we race towards Earth system collapse, trying to raise the alarm feels like being trapped behind a thick plate of glass. People can see our mouths opening and closing, but they struggle to hear what we are saying. As we frantically bang the glass, we look ever crazier. And feel it. The situation is genuinely maddening. I've been working on these issues since I was 22, and full of confidence and hope. I'm about to turn 59, and the confidence is turning to cold fear, the hope to horror. As manufactured indifference ensures that we remain unheard, it becomes ever harder to know how to hold it together. I cry most days now."[7]

We talk about the floods and droughts, the fires and the melting ice, evacuations and disaster relief. We talk about degrees Celsius and parts per million. Or at least some of us do. I'm only just learning to. But even fewer of us, I sense, are talking about how we feel about all this. We're afraid of climate change and perhaps equally afraid to feel how scared we are. We're furious at government inaction, but we don't want to be "an angry person." We feel helpless, and this is so uncomfortable that we don't think we can stand it. We're sad yet holding back, wary of drowning in this sadness; it might be bottomless. We don't want to cry most days. Nonetheless it is my hope—a strange thing to hope for—that this is changing and the time has come to tell each other how the climate crisis feels, how hard this is; not for the sake of wallowing, obviously, or even just for getting through the day (though sometimes just for getting through the day) but for the sake of motivation and transformation. My most honest conversations so far have showed me there's relief in giving up the charade that everything's okay and strength to be found in this deeper communion. Maybe we cry, but we don't have to cry alone. Maybe we crack each other up with some much-needed gallows humor. Maybe we're not crazy. Maybe it literally makes the difference, for some people, between wanting to kill themselves and wanting to go on living. Maybe, just maybe, we don't turn on each other as we face scarcity because maybe this crisis brings out the best in us, our capacity for tenderness, caring, and cooperation. Maybe, whatever ends up happening, we come together and take care of one another.

As Joanna says, "This is a birthing time as well as a dying time. It's going to take this kind of anguish to birth a new humanity, a new solidarity."[8]

IN AN ESSAY WRITTEN as a tribute to Joanna's work, Andy Fisher, a pioneer in the field of ecopsychology, observes that "every society shapes its members into a personality structure congruent with the continuation of that society. In an eco-destructive society, this means we are forced to repress or marginalize the wrenching pain we feel over the wasting of the earth in order to be the cheerful consumers of it."[9]

I take this as permission to not feel cheerful. Not only that, but is Andy suggesting that the opposite could be true? Maybe it is an eco-*constructive* act, or the beginning of one, at least, to just be honest and feel this pain. I also wonder, since he mentioned repression, if that isn't related to denial. It occurs to me that I can have more compassion for climate-change deniers or climate-change minimizers—the ones going about business as usual as well as the decreasing numbers of ones explicitly arguing that it doesn't exist—when I consider that denial is the first of the Elisabeth Kübler-Ross stages of grieving.

MORAL OUTRAGE

I read that since the signing of the Paris Agreement in 2016, "more carbon has been added to the atmosphere than in the entire history of humanity through the end of World War II."[10] Global greenhouse gas emissions have *increased*, and not a single country has reduced emissions in these critical six years.[11] How does that make you feel? It makes me want to scream.

Kübler-Ross stage 2 is anger. Remember when Donald Trump, as the president of the United States and perfectly in character as pot notorious for calling kettles black, tweeted that Greta should "work on her Anger Management problem"? I don't accept the premise that she has an anger "problem," now or then. The young woman I met in 2021 in conversation with the Dalai Lama was thoughtful, polite, kind, and so self-aware that, as I already mentioned, she winked at the tone of her 2019 "How dare you . . ." speech to the UN, calling it "very dramatic." The way I see it, the anger in her speeches, when it's there, is both perfectly appropriate to the moment and skillfully expressed. Rewatch that speech and you'll see a girl harnessing the power of her anger, a young woman who has metabolized raw feeling into right action—in Joanna's terms, a person who has found a wholesome and healthy way to suffer with her world. (Not to mention, Greta effectively rested her case by waiting a year to answer, then, when Donald Trump began his very public tantrum after losing the 2020 election, she simply, cheekily tweeted his words back: "So ridiculous. Donald must work on his Anger Management problem, then go to a good old-fashioned movie with a friend! Chill, Donald, Chill!")[12]

My friend the Zen priest and anthropologist Roshi Joan Halifax talks about moral suffering, which she defines as "the very harm we experience in relation to actions that transgress our tenets of basic goodness" and breaks down into four main kinds.

Moral distress, she says, is the "anguish that arises in the mind, in the body, or in our relationships when we are aware of a moral problem that we feel we have a responsibility to address, and we might have a determined remedy, but we're unable to act on it because of internal or external constraints." Directly or indirectly, wittingly or unwittingly, in our lack of action we become participants in the moral wrongdoing that is the very source of our distress.

The second form of moral suffering, according to Roshi Joan, is that of moral injury. She says it's usually associated with the military, but also with medicine, as when health-care systems force workers to act against their personal and even professional ethics. "Moral injury," says Roshi Joan, "is a psychological wound resulting from witnessing or participating in a morally transgressive act or actions. It's a toxic, festering mix of guilt, shame, and dread."

Moral outrage, by contrast, "is an externalized expression of indignation toward others who violate social norms. And it's a response that involves both anger and disgust." A common misunderstanding of Buddhist practice or even just mindfulness assumes that whatever chaos or injustice is happening around us, ideally you will find us with our eyes closed, sitting blissed out under a tree. This mistakes equanimity, a quality prized by Buddhists, with being neutral, unconcerned, disengaged. On the contrary, equanimity is a state that makes it possible to acknowledge our broken hearts and get curious about our darkest and most difficult thoughts and feelings without becoming overwhelmed or destructively reactive. As Roshi Joan puts it, "Equanimity is that state of mind that actually is able to include everything and sustain resilience, to have the ability to look deeply and to see when others are engaging in harm. It is not a flattening out but an opening up." As such, equanimity is not the opposite of moral engagement and action but the key.

Roshi Joan has a name for disengagement, and it's the fourth kind of moral suffering—moral apathy. She elaborates: "That's when we simply do not care to know, or when we are in denial about situations that cause harm. Moral apathy can be indifference that's based on privilege, can be indifference that's based on denial or addiction." She credits the writer James Baldwin with this last category, from watching Raoul Peck's documentary *I Am Not Your Negro*, which imagines in film a book Baldwin had planned to write, but he died before he had more than notes, about America told through the lives of Martin Luther King Jr., Malcolm X, and Medgar Evers. I watch the movie and find the part. It's black-and-white footage of Baldwin speaking calmly, precisely, to an interviewer: "I'm terrified at the moral apathy, the death of the heart, which is happening in my country. These people have deluded themselves for so long that they really don't think I'm human. I base this on their conduct, not on what they say."[13] Based on the conduct of industrialized societies since the first IPCC report, since we have known better, one could similarly conclude that we don't really think climate change is real. I remember what Vandana Shiva said about the exploitive logic of Industrial Growth Society being one and the same as the logic of slavery. James Baldwin's words and *I Am Not Your Negro* are light bursting and beaming from the heartbreaking darkness of racism.

Clearly, of all these forms of moral suffering, moral outrage is the one that is clear-eyed, sensitive to reality, *and* naturally leads to engagement, the one that makes a call to action out of our pain. Roshi Joan notes the psychologists Nancy Eisenberg and C. Daniel Batson, whose work has shown that some level of arousal is necessary for taking altruistic action, and reminds us that the Dalai Lama, too, speaks about how anger or moral outrage can initiate principled action. He's well aware that anger can go wrong, of course, but takes the Buddhist position that anger in itself is not necessarily negative. It is a fierce state of mind that, skillfully handled or channeled, can give us energy and courage. Particularly in the West, His Holiness says, Buddhists can be overly wary of anger. He challenges us to look deeper at moral outrage and see that what underlies it is compassion. Greta and her fellow activists are angry about the climate crisis because they *care* about people, animals, and landscapes—about one another and about future life on this planet. It's an appropriate response that brings edge and energy to the movement. No, we should not calm down.

So, moral outrage can motivate us to right wrongs and take action. I find, though, that in reckoning with human responsibility for climate change and environmental destruction and the moral suffering this inevitably entails, if I'm being honest, I have to rummage through the whole mixed bag that Roshi Joan describes: moral distress in the form of guilt and shame because I'm helpless to change the past; moral injury because functioning in the society I happened to be born into requires me to participate to a certain extent; moral outrage at the sluggish pace of change and lack of political will to take immediate, significant action; and moral apathy sometimes because I get distracted and it's easier to ignore the truth.

Working in West Germany in the 1980s, Joanna Macy met a former SS officer who told her, "It is easier for me to respect people who know they produced a Hitler than a nation so convinced of its innocence it imagines it could never produce one."[14] His was one example, she says, of Germans' willingness to "face their moral pain over their Nazi past." Joanna says not only was it remarkable to behold a society emerging from the silence and repression of the previous generation, daring to say out loud their deepest feelings of moral suffering, but that she was exceptionally creative when she was there, which she attributes to this atmosphere of moral courage.[15] With climate science came a loss of innocence, one that's still coming with every new study and every new report. We know better now. We know what we're doing, and we can effectively see into the future—not with perfect certainty but enough to break our hearts. What are we going to do about that? How can we change and how can we bear what we've done? With Greta as an example, or Germany for that matter, how might we channel our suffering?

"There's a term that we use in Zen that I really love," says Roshi Joan, "and it is *robai-shin*. And robai-shin is translated as 'grandmother's heart.' As we know, grandmothers are not pushovers. They have had the incredible challenge of being born into a woman's body. They've given birth. They've experienced loss. They've experienced brokenheartedness. And they have realized wisdom and tenderness." To have grandmother's heart, she says, is to recognize the "tremendous potential for positive transformation" in this moment, the "global rite of passage" that we are going through today. What does this take? Equanimity, she says, to hold our pain and outrage. And practice.

PRACTICING WITH HEARTBREAK

Kritee Kanko

Climate scientist, educator-activist, grief-ritual leader, and Zen priest

This practice came out of gatherings at Lama Foundation, led by my beloved grief ritual teacher Beth Garrigus. It is also heavily influenced by the powerful "Truth Mandala" ritual developed by the Buddhist eco-philosopher Joanna Macy. This ritual, which honors our grief, can be performed by an individual but is most effective when done in the presence of a small group of people that intentionally come together to bear witness to one another. Many of us have a hard time dropping into our feelings, but when even one person in the group opens up to their grief, it also encourages others to be vulnerable. Like every practice, the experience of this ritual deepens over time.

Preparation: Create an altar by putting images or other symbols of real or mythical ancestors, teachers, friends, and nonhuman relatives that you or your group feel gratitude for. This altar will act as a loving guardian and an alchemical vessel for holding and processing our grief and associated emotions. Organize these objects in a circle and divide it into four imaginary quadrants. In each quadrant there is an object: a stone, dry leaves, a thick wooden stick, and an empty bowl.

- The stone is for fear. A stone is how our heart feels when we're afraid: tight, frozen, and hard.
- The dry leaves represent our grief for what is happening to our world and for what has already fallen or been taken away from us.
- The wooden stick represents our anger and outrage.
- The empty bowl stands for our sense of confusion, uncertainty, and not knowing how to act or what to do in these times.

Instructions for the ritual:

- Meditate, chant, or sing for five minutes to arrive in the space and the present moment.
- Remember everyone who makes up the boundary on your altar. Allow yourself to feel connection and gratitude for at least fifteen to twenty minutes. Ask these guardian energies to help you/the group process the painful emotions.

- Then pick up the four objects (stone, leaves, stick, or bowl) one by one. If you are alone, allow yourself to feel emotions associated with one object before you go to the next object. If you are with others, you can stay silent and just hold the stone/leaves/stick/bowl to express your feelings, or you can speak about these feelings after making agreements about the length of the ritual (forty-five to sixty minutes).
- The body doesn't differentiate between grief that is due to personal or ancestral trauma, or cultural oppression due to race, gender, economy, or climate. So, you are invited to do two rounds: The first round is for honoring personal/ancestral/racial/gender-based trauma. The second round is for honoring pain associated with the climate crisis.
- Please breathe deeply and allow gentle movement of the body and eyes. Rock, swing, and/or sway gently as you listen to yourself and others. Tears, sounds to acknowledge others, wailing, yawning, or even hiccups are natural and okay.
- You can self-soothe yourself, but don't comfort others until the ritual is complete and unless clear agreements are in place.

At the end of the ritual, remind yourself (and others) that each symbolic object is like a coin with two sides. Grief exists only because we love. Anger exists because we want justice. Confusion and uncertainty are the ground from which new direction emerges. Fear speaks about our courage to face and name our fears.

Thank yourself and others (if any) and the guardian energies before you leave the ritual space.

METABOLIZING CLIMATE HEARTBREAK

My other friend Stephanie, Stephanie Higgs, also works in publishing (and cowrote this book). Years ago, she had a mentor at Random House, the editorial director Daniel Menaker. More recently, Stephanie tells me, Dan wrote a little book of poems while he was living and dying with pancreatic cancer, in which he struggles to make something beautiful out of his sickness, pain, and grief. As I was talking to her about this chapter, Stephanie shared a stanza from one of Dan's poems, called "Adjuvants," from this collection, called *Terminalia*. (She says he always did savor a well-placed four-thousand-dollar word, in conversation as much as on the page, maybe partly for the wow factor, certainly because he loved language, but also because he liked to teach. I know *adjuvant* because of my nurse's training in cancer care—it means a secondary therapy that helps the efficacy or potency of the primary one—but Stephanie says she had to look it up.) The poem describes friends coming over to visit during a two-week pause in his chemotherapy treatments; and in the last lines, after they leave, how he weeps "with joy and its twin brother mourning."[16]

This is what I'm trying to do with this chapter, too. I'm hoping that together we can hold the space of heartbreak long enough to find the joy in it, the love in it, to let a little light in and glimpse what we might create, not in spite of our broken hearts but because of them. We grieve because we love.

"We have a lot to grieve," says Camille Barton. Camille is a Berlin-based artist, educator, and self-described "embodiment researcher" who created something they call a "grief toolkit."[17] (Camille uses they/them pronouns.) It sounds practical, and in collaboration with the Global Environments Network they offer it freely online. I'm curious to check it out. "We have a lot to grieve," Camille says, "yet in the West, we are living in incredibly grief-phobic societies. Grief is taboo. There isn't much public space available to be with our loss or feelings that are not 'productive.'" They point out how we are expected to deal with our grief privately, quickly, and quietly, perhaps with psychotherapy if we can afford it; and that as far as Industrial Growth Society is concerned, the point of grieving is to "'get back to normal' and become productive members of society again." From researching other approaches, particularly in Africa, Camille assembled the toolkit to "challenge this logic of private, secret grief."

"Grief work can be a way to acknowledge and feel into what needs to shift as well as what you love and wish to be in a deeper relationship with," according to Camille. "It seems that grief may be important medicine at this moment where there is an opportunity to consider how we wish to live on this planet and with each other." I do wish to consider these things! So, what's in the kit? It's got four main compartments. The first contains "embodiment tools," which we can use, like Willa Blythe Baker's grounding exercise later in this chapter, to reinhabit our body when we are disconnected from the present and/or too much in our head. These simple tools include, for example, placing one hand on your chest and the other on your belly; or one hand on your forehead and the other on the back of your neck. Second, the kit offers "personal practices." Noticing that more than one of these involve writing, I think of Daniel Menaker working on those poems through chemo, as he was saying goodbye to everyone and everything he loved. One personal practice is simply called "Pillow Screaming." There's a ritual for dancing our grief, fire- or water-based rituals to help us let go, and moaning and rocking to "support grief to flow."

The third part of the kit shows us different ways to grieve together, which reminds me of what Kritee Kanko says about the power of people coming together to bear witness to one another. Camille's group practices come with helpful instructions for creating mutually supportive spaces, for example a "Sharing Circle" ("Decide on the amount of time you will hold space for each other. For example, you might choose to close after one hour or after everyone has shared three times. . . . Decide on any ground rules or shared agreements: i.e., confidentiality, not interrupting each other, no advice unless requested or other rules that feel supportive.") or a "Candle Vigil" ("If your vigil is part of a demonstration or march, you might like to walk with lit candles and images, then share memories or speeches at a set destination at the end."). I like the sound of a group practice called "A Grief Date with the Forest or Ocean."

Finally, the toolkit offers some "integration practices," which Camille says help us to reflect on a given ritual and "signal" to our body, as we are about to exit the ritual space, that it's time to shift. "At its most basic," says Camille, "ritual is anything that is practiced with an intention or a specific goal or purpose." They point out that in Industrial Growth Society, we regularly perform rituals "that support resource extraction and relationships based on domination." By contrast, Camille suggests rituals that "create space to practice or explore ways of being" that support nurturing, egalitarian relationships with each other, all beings, and the Earth, or in Camille's words, that "sustain life."

While the rituals and exercises in this toolkit may seem simple, the author acknowledges that we, especially Westerners who are out of practice, may in practice find them hard. "Please be gentle with yourself," says Camille, and I think, *We don't say this to each other often enough.* So, I'll say it again: please be gentle with yourself.

"There are ways through," agrees Willa Blythe Baker, "ways through the heartbreak into heart opening and action." Willa is a teacher (lama) of Tibetan Buddhism with a PhD in religion, and another dear friend and mentor I welcome to this conversation.

There's the tradition of *lojong*, for example, radically counterintuitive heart-and-mind training practices where you "want the
difficult," Willa says. Instead of resisting it, you invite it in. "We
choose to engage with challenge in our life, including the sorrow,
the despair, the grief, all of that, because otherwise you couldn't
possibly know other people's sorrow, grief, despair." This way, the
difficulty becomes a "bridge to compassion for other people, then
solidarity with other people, and then action on behalf of other
people." Willa calls lojong and other contemplative practices for
working with our hardest thoughts and feelings a kind of "metabolizing process," where (to mix a metaphor) "every disturbing
emotion is a bridge."[18]

"One of the feelings that arises for many of us," Willa acknowledges, "when we lean into the truth of climate change, is
despair. The facts themselves, the science itself, the reports that
we read about in the media, the research—which is all important
to listen and attend to—can bring up fear, urgency, powerlessness, sadness, overwhelm, and despair."[19] Willa says these feelings are so uncomfortable that we tend to think we either have to
avoid them or solve them. But she suggests a different way to be
with them. I've done guided meditations with Willa where she
encourages us to *befriend* despair. Talk about counterintuitive.
But not, I found, impossible when she helps us see two things.
First, that despair, like any other feeling, "is not a static state. It is a
flow state." It comes and goes and comes and goes and shapeshifts
and changes in tone. Second, Willa guides us to recognize that
despair is "an outpouring based on our encounter with the truth
that something we deeply love is at risk. It's our love that brings
us toward this issue, and it's also our love that causes our hearts
to break. Despair is a symptom of our tenderness for the natural
world, for the plants and the animals. It is a response to the realization that life is impermanent. And the life of the planet is frag-

ile. Despair is the response to overwhelm—to the truth that we cannot control collective greed and delusion, to the truth that our daily human actions contribute to the problem due to systemic collective agreements that were made long before we were born. In that light, for an organism experiencing overwhelm, despair is one natural response. It's not an aberration, or a failure of will." When we acknowledge the love in our grief and despair, Willa is saying, we can access this feeling of tenderness that is itself a kind of light. When we hold the truth of climate change and our feelings about it in a certain way, rather than dissolve into them, we can metabolize them into caring. And caring motivates action.

I tell Willa that I appreciate the bodily connotations of "metabolization," and I'm aware she chose this metaphor carefully. The author of a book called *The Wakeful Body: Somatic Mindfulness as a Path to Freedom*, Willa knows as well as anyone that what we're talking about here—these ways through—cannot happen only in the head. In fact, she says when she is working with people who are new to such practices, who are perhaps avoiding talking about climate change or telling themselves everything will be fine, maybe even resistant to feeling any feelings at all, the body is the starting place that she suggests. She'll give them something specific to do, like feeling the ground beneath them, and this, she says, "has much more potential to teach that person than anything I could say." Reflecting some more on metabolic processes as I learned them in nursing school, I realize the analogy is perfect. When we metabolize food, it nourishes us; it actually gives us life and lets us keep going. So, too, with the fuel that we make out of our heartbreak: a way to keep going through compassion and action.

FINDING GROUND

Willa Blythe Baker

Buddhist author, translator, and founding teacher and spiritual codirector of Natural Dharma Fellowship and Wonderwell Mountain Refuge

Ours is a terrestrial body. Like the big Earth is pulled toward the sun, the small earth of the body is pulled toward this planet, and while our journey is not an orbit, it is a grounding.

Everything we do is predicated on the existence of gravity, and yet how often do we notice it? It is a powerful and ubiquitous force, and yet how often do we consider its power? In cultivating a practice of somatic mindfulness, attention to gravity is important for one simple reason: *gravity grounds us.*

Notice the weight of your body. Where do you most feel the weight of your body? Let your awareness settle into the place where your seat, legs, and/or feet rest against the cushion or chair you are sitting on. With curiosity, explore this feeling of pressure and groundedness, where your body contacts the earth, floor, chair, or cushion.

At this location, you can actually experience gravity, not as an idea but as a feeling. You can feel your body's attraction to the earth. Can you feel your body's natural groundedness and stability? Let your attention gather there, dissolve there.

When your mind becomes restless or preoccupied, let your body's groundedness draw your mind back, like a magnet draws iron filings. Notice how your body teaches your mind to be stable and still. You might say the body is capable of exerting its own gravitational pull on the mind.

When you notice yourself getting drawn into the mind's dramas, offer yourself this simple instruction: "Ground."

Adapted from *The Wakeful Body: Somatic Mindfulness as a Path to Freedom*[20]

HOW SHOULD WE LIVE NOW?

One thing we should establish straight away," said Ralphy, "are the rules. Rule No. 1: no murder." At this, the girls broke into a chorus of laughter.

"Murder?!" asked Erica.

"Literally the only thing we're trying to do on this island is not die. Why—" but Sam couldn't finish her sentence. She was laughing too hard.

"Why would we murder anyone?" her twin finished. Peals of laughter rang out from the beach and carried across the water.

"Stop!" yelped Ralphy. "Stop! I'm seriously going to pee myself!"

Jackie wiped tears from her eyes. "Oh, my God," she said. "I needed this."

Later, when the girls were ragged and hungry, all any of them had to say to cheer the group up was "murder," and that would set them all off giggling again.[21]

Literally the only thing we're trying to do on this planet is not die, to paraphrase the humor writer Riane Konc's Sam character in her all-female reimagining of *Lord of the Flies*. And at least in terms of average life expectancies and overall population growth, to date humans have been wildly successful at not dying. "We have proven our superiority at figuring things out and removing obstacles to our desires," says Catherine Ingram in the voice of dominant human societies. "We killed off most of the large wild mammals and most of the Indigenous peoples in order to take their lands. We bent nature to our will, paved over its forests and grasslands, rerouted and dammed its rivers, dug up what the journalist Thom Hartmann calls its 'ancient sunlight,' and burned that dead creature goo into the atmosphere so that our vehicles could motor us around on land, sea, and air." Uncontroversial so far, but there's a catch that we are finally grasping after so many millennia of evolution and, by many measures, of seeming to have won at life; a catch captured by a favorite saying of Catherine's: "Nature bats last."[22]

Note that the reason Riane's piece works is in the "we." Change it to "literally the one thing *I'm* doing is trying not to die" and it falls apart (and stops being funny). It becomes the original *Lord of the Flies*. A "no murder" rule is only ridiculous with the recognition that we are on this island, or this planet, together and that my not dying depends a lot on you not killing me and vice versa. In this sense, it's as effective an illustration of the law of interdependence as Indra's Net. Many people, Catherine Ingram among them, talk about how humans will become increasingly selfish and terrible to each other under the pressures of climate change. But there's an alternative, as Riane suggests, so obvious as to be funny that it needs being said: to not be terrible to each other. To not kill each other or the ecosystems on which our lives depend.

The thing we're trying to do is not die. Yet we're all going to die. These things are equally true.

Death is certain;
The timing of death is uncertain;
How should we live now?

For 2,500 years, Buddhists have reflected on these lines, as true now as they ever were. If this sounds morbid, it's actually not. The Buddhist scholar Stephen Batchelor relishes the irony: "Only when we lose the use of something taken for granted (whether the telephone or an eye) are we jolted into a recognition of its value. When the phone is fixed, the bandage removed from the eye, we briefly rejoice in their restoration but swiftly forget them again. In taking them for granted, we cease to be conscious of them. In taking life for granted, we likewise fail to notice it. (To the extent that we get bored and long for something exciting to happen.) By meditating on death, we paradoxically become conscious of life. How extraordinary it is to be here at all."[23] (If you would like some guidance for this death meditation, Stephen takes readers through it line by line in his book *Buddhism without Beliefs*.) It's true that as hard as my awakening to the climate crisis has been, it has also made me fall in love all over again with life. Not only life in general but specific life. That tree. This friend. That time sitting on my screen porch when the light itself seemed to say, "Good morning." My brother who was diagnosed with cancer while I was working on this book.

Partly in response to what Catherine Ingram says about facing extinction, Stephen began to consider those lines anew, in terms of the human species.[24] *Extinction is certain . . .*[25] He concurs that human extinction is certain, but he is less certain than Catherine that it will happen soon. *The timing of extinction is uncertain . . .* "Either the human species will evolve into a form of life that we cannot now even imagine, or, if we manage to survive in a more or less humanoid form, we will be wiped out when the sun becomes too hot to sustain life on Earth in around a billion years."[26] Or we will make our planet unlivable for us much sooner. Whichever it turns out to be, Stephen wants to emphasize the uncertainty as much as the certainty. We don't know, he says, and there could be something beautiful and meaningful in this not knowing, for however long we have, if we choose. *How should we live now?*

Joanna Macy says her way of dealing with heartbreak "is to look it straight in the eye and say 'there you are again.'"[27] This reminds me that our feelings aren't linear. We talk about the Kübler-Ross "stages" as if we arrive at some final endpoint where we can say, "That's it, I'm done grieving." I've heard Steve Leder, the senior rabbi of the Wilshire Boulevard Temple in Los Angeles and author of a book about death and grief called *The Beauty of What Remains*,[28] say that when he counsels grieving families, he doesn't tell them, "It won't always hurt this much," because that's not true. The truth for people who've lost someone they love, he says, is that it won't always hurt this often.[29] Of course, our grief for Earth is different. It may always hurt this often. Our hearts break over and over at some fresh hell of climate or environment news or lived experience every day, and the losses we're talking about are ongoing—past, present, and future. At any rate, we will not be done grieving. At the same time, we will not be done loving or done trying as long as we live.

6

WONDERMENT

A Present We Can Love, a Future We Can Imagine

HEARTBROKEN IN THE MONTHS after the 2016 American presidential election, Jenny Odell found herself returning again and again to a public rose garden in her neighborhood in Oakland.[1] At first she didn't have a plan, didn't really know why she was there—she recalls just sitting with a "thousand-yard stare,"[2] but in retrospect she describes some kind of life-preserving instinct nudging her out of her apartment to that garden. She was, she says, "like a deer going to a salt lick or a goat going to the top of a hill,"[3] following some inner impulse to get away from the obsessive doomscrolling she had been doing on her phone. In a talk she later gave about the meaning of her time in the rose garden, which turned into the first chapter of a book, *How to Do Nothing*, Jenny reclaims the idea of stopping to smell the flowers from the realm of cliché: "I know that in the months after the election, a lot of people found themselves searching for this thing called 'truth,' but what I also felt to be missing was just reality, something I could point to after all of this and say, This is really real."[4]

Jenny's being modest when she says "just" reality and doing "nothing," because as she sat there wondering at the plants and animals all around her and the human volunteers making the effort to take care of them, the scene inspired some keen insights and restorative feelings. She tells us how, when she realized the saving grace of finding her place—her physical, living, breathing place—as a human animal in nature, "I grabbed on to it like a life raft, and I haven't let go."

"*This* is real," says Jenny. "Your eyes reading this text, your hands, your breath, the time of day, the place where you are reading this—these things are real. I'm real too. I am not an avatar, a set of preferences, or some smooth cognitive force; I'm lumpy and porous, I'm an animal, I hurt sometimes, and I'm different one day to the next. I hear, see, and smell things in a world where others also hear, see, and smell me. And it takes a break to remember that: a break to do nothing, to just listen, to remember in the deepest sense what, when, and where we are."[5]

If you've ever cared about someone who was struggling with a substance problem—and I have—there's another bit of wisdom from that experience that we can apply here. *This* is not only real, it is repugnant and transcendent and scary and encouraging and frustrating and outrageous and heartrending and so, so beloved all at the same time. Ugliness and beauty, fear and hope, frustration and anger, and compassion and love can all be true. The light shines in the darkness. You learn to live with these contradictions because you need the light to see. In short, the bad doesn't cancel out the good.

As terrible as it is to consider the destruction we've wreaked on this planet and notwithstanding the wretched possibility that it's too late to turn climate change around, Jenny and the other people we'll speak to in this chapter remind me that heartbreak isn't the only way the light gets in. Talking about the importance of awe—or "wonderment," as Matthieu Ricard, the Buddhist monk and photographer we met in chapter 4, charmingly calls it—in our relationship with nature (human nature included), I marvel at the power of wonder not only to shine in the darkness and make us feel better but also to stir our will to action.

As his name suggests, Matthieu's first language is French. He says even after decades of speaking English, he still finds it nearly impossible to get his mouth around the word *awe*, the word used by most researchers who study the topic. As he puts it, *"Wonderment* easier for French guy raised in the Himalayas." I have adopted his word and made it the title of this chapter because to me, *wonder* contains everything from a wide-eyed, childlike curious heart ("full of wonder") to a wise open mind ("I wonder..."). And while both *awe* and *wonderment* may point to the same lovely feeling, the extra syllables of *wonderment* that make it easier for Matthieu to pronounce also make it easier for me to savor since they take that much longer to say or type.

Matthieu likes to track studies and applications of wonderment. "Two years ago," he says, "I heard of a campaign by the Greens in Germany who put all kinds of beautiful images of nature throughout the city, big images." The effect was that "everybody got really inspired and motivated. Maybe we could use wonderment as a drive?"

"There are interesting studies, scientific studies done by our friend Dacher Keltner," Matthieu continues. Dacher—a psychologist at UC Berkeley and a Mind & Life colleague who has been studying awe, to use his word, for decades—and his colleagues have shown that in addition to being a source of motivation, wonderment increases altruism, because it is an opening of the senses to a reality bigger and greater than ourselves. (It figures, then, that humility like Jenny's would go with it.) A new study Dacher was involved in, for example, measured the benefits of what the researchers call "awe walks." Participants who took brief, weekly walks during which they made a point of staying in the present moment and taking in their surroundings with fresh eyes felt more buoyant and hopeful in general, not only on the walks, than participants in the study who did not.[6]

There are different kinds of wonderment as well as different effects, Matthieu says, offering the example of a teacher he reveres. "I was so fortunate to spend thirteen years at his feet and studying with him. So that's wonderment about the incredible potential of human nature that can bring to maturity such people of unconditional love and bottomless wisdom." This kind of wonderment can inspire us to follow such a teacher's example. Sharing a photo of Kangyur Rinpoche, Matthieu tells us he became his first teacher when they met in Darjeeling, India, in 1967, where the Tibetan lama was exiled. Matthieu points to the eyes, sparking with love and wonderment, "a perfect Duchenne smile." "You can try it later," he encourages, resuming the role of the teacher himself. "So, there's a wonderment about the best thing that human beings can accomplish," he goes on, "then you have wonderment about human beings in general. Let's not underestimate the banality of goodness." The phrase intrigues me. Matthieu says it means that most of the time, most of us eight billion human beings behave civilly with each other, and that's why, for example, upon leaving a train, a plane, or any other public space where people were decent, we don't say, "Wow, nobody had a fight," because it's normal. What's abnormal, Matthieu says, is what is always on the news because it's a deviance, a potential danger, adding, "And there's always something wrong happening, something terrible in the world." Too much news and we fall into "wicked world syndrome," like Jenny's doomscrolling.

"So, let's take back confidence in the beauty of the basic goodness of human nature," Matthieu concludes. This, he says, is what he's tried to do with his photography, and he shows us some more examples: an old Tibetan man wrapped in a purple coat with furry trim, bending toward the camera mid-laughter, missing a few teeth and smiling the widest smile; a young girl in a bamboo classroom in Nepal, waving or dancing or clapping or all of the above, "so glad to have an education," Matthieu explains. He founded the organization, Karuna-Shechen, that built her school.

Now we're looking at a monk sitting cross-legged on flat, rocky, almost otherworldly ground, facing a single mountain. Because the monk's back is to the camera, I can't tell whether he's meditating or having a conversation with the mountain. The mountain looks patient. The sun catches the monk's red robe. As Matthieu captions this one, "We need also to wonder when we see the beauty of the world." He's getting at another kind of wonderment, as a source of respect. "You don't want to destroy what gives you wonder, something so vast, so unique, that took so many, many million years to form." I think of Wangari Maathai's tree by the stream in Kenya, surviving as long as people believed it to be divine. "If you respect something," Matthieu says, "then you will naturally care for that thing, and if the thing is threatened, then care will naturally translate into action." There's the motivating effect of wonderment, again.

Next slide. "Oh, this still lake at about twelve thousand feet in Bhutan with a twenty-one-thousand-foot mountain at six in the morning." And with this photograph, Matthieu gives us a perfect description of how wonderment feels: "At this extraordinary place it seems that the meditation is as much outside as inside. And even though you are alone there, you feel a sense of connection, of interdependence. Many studies have shown that loneliness is terrible for mental and even physical health. But wonderment brings you that sense of connection with the whole biosphere, all sentient beings, the universe, the big sky, and you naturally want to take care of them."

NATURE SHOWS

My friend Thupten Jinpa, the Dalai Lama's English translator, says His Holiness doesn't watch much TV (as you might have guessed) but that he really likes nature programs by the English broadcaster and naturalist David Attenborough—it makes me laugh when Jinpa goes so far as to call the Dalai Lama "a big fan." Jinpa remembers renting a car in New Delhi with a friend and hauling a massive television set, back in the days before flat screens, to the Dalai Lama's residence about a ten-hour drive north, in Dharamshala. They brought a whole DVD set of a BBC nature series narrated by David Attenborough to go with it. Jinpa says the Dalai Lama appreciates the same things as the rest of us about these shows, especially the sense of wonderment that they inspire. "Through color, space, expansive vistas, tremendous diversity," in Jinpa's words, "the good nature programs really give you that sense of the grandeur of it all." He stops as if that's all he has to say, then has another thought. "The modern attitude toward nature is purely as a resource to extract from. We relate to nature in utilitarian terms." Whereas, Jinpa says, the way nature shows invite us to see it, nature becomes something to wonder at and wonder about.

In this part of the book where we're talking about will to action, what Jinpa said fits right along with Jenny Odell and Matthieu Ricard. In different ways, they each get me thinking about different definitions of action—action as slowing down, as attuning to nature, as wonderment, as respect. By relating to nature in utilitarian and hubristic terms, Industrial Growth Society confuses doing with production. Famously, in 2019, Greta crossed the Atlantic Ocean in a fossil-fuel-free sailboat to attend UN climate talks in North and South America,[7] to make a point about the world's existing transportation infrastructure being out of step with the needs of the planet. Landing in New York City, she said she would miss the kind of action she had gotten used to on the boat. "To just sit, literally sit for hours, and just stare at the ocean not doing anything. That was great. And I'm going to miss that a lot. And of course, to be in this wilderness, the ocean, and to see the beauty of it."[8] In Industrial Growth Society, being productive means having something new to show for our time;

in particular, something that's worth money. Industrial Growth Society says, "Look at us! Look what we made out of nature!" Wonderment is the opposite. Wonderment watches and listens, notices how things are, lets things be, takes care, does no harm. Wonderment doesn't think we know everything; doesn't have to own everything; and it recognizes that sometimes, the best something we can do is nothing.

A downside to the nature-show gaze, though, is that it can put us on the outside looking in, and in Industrial Growth Society, as we've talked about, a sense of separation from nature is very much part of the problem. Watching a flock of bar-headed geese fly over the Himalayas (one of my favorite parts of the *Planet Earth* series), do we remember our place in nature—the melting ice on the Tibetan Plateau, the people who live on the ground, the plane with the camera that got the shot, the fact that humans burning fossil fuels halfway around the world affects these birds—or are we watching to forget? The show can cut away to David Attenborough on a balcony overlooking a city talking about human impacts, but when we get back to "nature," humans are conspicuously not in the scene. On the one hand, such scenes elicit wonderment in the best sense, because we need to remember we're not the center of the universe and it's not always about us. On the other hand, wonderment in what we might call the safari sense creates the illusion that we can sit comfortably in the audience while we enjoy the show, a spectacle of "untouched" nature "out there."

Carl Hall, with his company West One International, is the distributor of the film *Earth Emergency*. He wanted to help with the film project partly because, he says, he felt guilty about his career making nature films. He had come to believe that they are lies that make us think nature's doing fine, habitats are intact, lions freely roam the plains as always. He tells a story about a film shoot of water buffalo at their watering place, where if you get there at the right time of year, you're likely to see a water buffalo being eaten by a crocodile. Classic, though "red in tooth and claw" scenes have never been my favorite. But this time, the British camera operator turned around mid-scene, revealing a hundred or so humans who were also there with their cameras filming the same thing. That was the whole truth of it, and Carl feels that the industry standard practice of leaving people out of it is lying. (The other reason Carl got involved was because his daughter, who is close to Greta's age, said, "Dad, you have to do this.") A producer of the Climate Emergency: Feedback Loops films says this came up at a conference of science producers that they had attended, too. They remember one speaker saying that for years, nature films have had to hide the reality of climate change or sneak it in somehow without turning off the audience; like hiding a little spinach in the dessert, you sugarcoat it, the speaker said. But he also made the point that people are more open to the whole truth now.

Some years ago, Mind & Life hosted a conference in Botswana. When the event was over, I went with some other attendees on a safari. We got up very early to see elephants, which have always been my animal. At first there weren't many other people around, and we got quite close to the elephants. You have to be quiet. For a short time, I had the feeling I'd come for, of wonderment and communion. But with the morning sun came the safari crowds, the wealthiest of them in helicopters. I became acutely aware that we were part of the scene as the scene became . . . less quiet. Now I was watching the elephants respond to the humans, and it was plain to me that the elephants were unsettled. We were intruding, uninvited, on their land. I grew up watching *Wild Kingdom* and reading *National Geographic*, falling in love with the images of wild animals and fantasizing about going to Africa. But my safari experience convinced me that we shouldn't be there, crowding around them and disturbing their peace. The path from wonderment to respect that Matthieu was talking about should lead us somewhere else and with a softer step, say in the direction of discovering wonders closer to home and protecting natural habitats.

EVERYDAY WONDERMENT

Talking to climate scientists reminds me that wonderment is part of their path. The sea ice specialist Don Perovich wants to tell us, for example, about working in the Arctic. "I wish that everybody could experience it," he says, although he and I and my elephants would agree it's better that not everybody *can* go. Let Don paint us a picture instead. "I've walked on frozen oceans under midnight suns. I've been up on the ice in November when it was so cold and so quiet you could hear your breath freeze, and above you are the dancing curtains of light of the aurora borealis," he says, with "creatures beyond belief, green blobs of waterfall phytoplankton, polar bears, walruses." Don says that his wondering, in graduate school, about how ice melts gave him an opportunity—"maybe an excuse"—to learn to scuba dive. "I remember one time I was diving, swimming along the side of the floe, and saw something out of the corner of my eye, really quickly. I kept on swimming. Then I saw it someplace else, just like a flicker, when you see things but don't really see them. I kept on going, and I looked down, and right underneath me, just a few feet away, was a seal, swimming upside down, just looking at me. It would flick its tail and be gone, and then it would reappear. I don't know who was more curious about the other, me or the seal." Don stayed with the moment until it was "so cold I couldn't keep my regulator in my mouth anymore," and the seal swam on for other adventures. He tells this story in answer to a question about how he not only keeps doing his work but keeps loving it, knowing what he knows about what we're doing to the Arctic. Wonderment keeps him going back.

Mind & Life, like many organizations these days, is doing a lot less travel for business than we used to. Unlike Don's job requiring him to show up sometimes in the actual Arctic, we've come to see how much of the travel we were doing wasn't necessary to our work; at the same time, we realized it was harming the planet. I'm also facing the fact that far-flung vacations on a whim are no longer an option, which is to say today I wouldn't fly across an ocean to go on a safari even if the elephants didn't mind. What do we do if we can't justify flying somewhere different from our ordinary lives to get a hit of wonderment? I imagine many people who've had that luxury fear the answer must be to live smaller, sadder, less wonderful lives. A related question is what have people who couldn't afford to go on exotic vacations anyway always done? I think it's the same answer for both questions: With a different kind of attention—curious, trusting, open, and loving—we can find the wonders of right where we are. Jenny Odell helps me see what wonderment looks like in the Age of Enough.

We heard about Jenny's love for a nearby rose garden, but she can also find wonderment on a city bus—or rather, since she's in the Bay Area, a streetcar. "Once I accepted the fact that each face I looked at (and I tried to look at each of them) was associated with an entire life—of birth, of childhood, of dreams and disappointments, of a universe of anxieties, hopes, grudges, and regrets totally distinct from mine—this slow scene became almost impossibly absorbing. As [the painter David] Hockney said: 'There's a lot to look at.'"[9] Jenny will tell you about making friends with two local crows, whom she encouraged to visit by placing a peanut each day on her apartment balcony. They quickly became regulars, and she gave them names: Crow and Crowson, since they are parent and child. As she describes, "I soon discovered that Crow and Crowson preferred it when I threw peanuts off the balcony so they could do fancy dives off the telephone line. They'd do twists, barrel rolls, and loops, which I made slow-motion videos of with the obsessiveness of a proud parent."[10] She started paying more attention to the crows after being curious enough to learn how smart they are—capable, for example, of recognizing human faces, which she took as an invitation to start a relationship. "Sometimes they wouldn't want any more peanuts," Jenny says, "and would just sit there and stare at me. One time Crowson followed me halfway down the street."

Jenny walks around her neighborhood with a jeweler's loupe in order to look at things magnified ten times.[11] Already a dedicated birdwatcher and flora appreciator, she says this simple (and inexpensive) gadget has opened her eyes to the everyday world around her in new and amazing ways. "You have to get really close to something with the loupe. And then all of a sudden it pops into focus and it's like this plant that you thought was smooth, it's really hairy or something, you know?" Receiving the pocket-sized high-power magnifying glass as a gift from a thoughtful friend who noticed how much she liked his, she thought, "I'll never be bored again." Which has been said about the smartphone, but Jenny says the difference is that "it fills me with wonder instead of dread."

On the other side of the country in Brooklyn, the meditation teacher and writer Sebene Selassie also talks of seeing the extraordinary in ordinary urban places. Like Jenny, Sebene ("like 'Seven-A' with a *b*," she says, "instead of a *v*") can find wonderment on a bus or, in her case, a subway. "Sometimes I like just to ride the subway," she says, by which she means taking the podcast out of her ears or putting down her book to be present and just ride.[12] "Over my nineteen years in this city, I have seen innumerable expressions of humanity on the subway—many beautiful, some brutal, some disturbing, many heartbreaking. I once saw a man punch a young woman in the face—a complete stranger, for no reason; he kept walking. Homeless people are ignored, and shamed, and dismissed every day (sometimes, I'm afraid to admit, by me)." Anyone who has ridden a New York City subway knows they aren't always clean or peaceful; Sebene's acknowledgment of that ugliness and suffering makes the wonderment more believable. "I have also seen countless people of every demographic help strangers with money, food, luggage, strollers, and directions," she says. "I have witnessed collective head bopping, laughter, compassion, and amazement. I hear every language and accent imaginable. I have watched acrobats, tap dancers, tango, break dancers, and musicians from every part of the world. It's like a secret underworld where all of humanity meets."

Sebene allows that sometimes she gets frustrated with the city. Sometimes, "I want it to be the country and to have more space and stillness. I want to be 'in nature'—as if there is anything that is not 'natural' in this universe." She is not beyond fantasizing sometimes about moving upstate, she says. And I will continue to fantasize about meeting animals in the wild and flying to places I've never been. But I welcome her reminder that "sometimes you just ride the subway." And sometimes it is wonderful.

Every spiritual tradition, including Buddhism, tells us that we don't have to leave wonderment to accident and happenstance, either. We can create holy days (the original meaning of holidays) or holy moments anytime with ritual and intention that shift our consciousness, heighten awareness, and get us out of habitual, business-as-usual mode.

WHAT IT MEANS TO SACRIFICE

from a sermon by *Steve Leder*

Senior rabbi of Wilshire Boulevard Temple, Los Angeles, California

Most people think of sacrifice as a net loss; giving something away that is precious. But the Hebrew word for *sacrifice* implies just the opposite. It means to "come close" or "draw near." From the Torah's perspective, when we give the right things up, we end up feeling closer to the places and people that matter so much more.

Not for a moment would I suggest the lessons the COVID-19 pandemic has come to teach us are worth the loss of life, the devastation to our economy, or the fear we are enduring; but neither are they worthless. Whether we realize it or not, despite the awful reason, we are creating something quite beautiful and hopefully lasting all over the world. We call this creating *via negationis*—by the negative. In other words, when we stop certain behaviors or remove certain things and even certain people from our lives, very often something quite unexpectedly beautiful emerges in their place.

Creating by ceasing to create is the essence of Shabbat, whose sanctity depends almost entirely on the things we stop doing. It is also the essence of Passover which is soon upon us. Think of removing *chametz* (leaven) from our homes and our diets as not only a physical act but also as a metaphor; the spring cleaning of our souls. At no time in my life have I felt the metaphorical power of removing the chametz of my life than during this quarantine.

Must we have so many suits, ties, shoes, purses, and outfits to wear? How many meetings do we really need to attend? Did we need to spend so much time away from home for so many years? How much do we have to spoil the earth by driving and buying, driving and buying, driving and buying? Like Shabbat, like Pesach, like all true sacrifices and then some, COVID has stripped away a lot of nonsense from our lives and in that stripping away something beautiful has emerged; a knowing that we were better meant to be home, closer to God, nature, and the people we love most. *Shabbat shalom . . .*

LOS ANGELES, 2020

Rebecca Solnit calls mobility "one of the fetishes of modernism and technology." Referencing the poet and environmental activist Gary Snyder, she says the most radical thing we can do is "stay home." I'm so glad to hear her state it in such emphatic and galvanizing terms, since I've been feeling increasingly strongly about making this sacrifice—which is to say increasingly interested in what we might stand to gain from it; not just the benefit to the planet of burning less fossil fuel to take us places but the personal and social possibilities of staying where we are and developing our sense of place. It's a big part of why I wanted to talk about wonderment, or as Rebecca puts it, "how to overcome the tyranny of the quantifiable of, like, 'I went to Burma and I went to Singapore and I went skiing here and I went shopping there.'" Rather, she wants us to wonder "how to make staying home exciting and rich and rewarding and adventurous."[13]

"The vast majority of us are going to feel much better once we slow down and live more locally," Greta and her family say in the book they wrote together, "safe in the knowledge that our children will get the chance to develop the inventions and solutions that we haven't managed to invent ourselves. The vast majority of us are going to feel much better when entire countries are given a chance to live instead of us forever being on our way to the next big city, the next trip, the next airport, the next whatever. The world gets bigger the slower we travel."[14] I think about how big the Atlantic Ocean must feel when you cross it in a sailboat, as Greta did.

THE PERFECTLY RIPE STRAWBERRY RIGHT IN FRONT OF US

Much of the time, when we are in decent health, we take our lives for granted, planning, expecting, worrying, and forgetting to appreciate the miraculous unlikeliness of each moment—living as if we'll never die. As Industrial Growth Society has amply demonstrated, we can be like this with our planet, too, taking life as we know it on Earth for granted as long as it seemed like it would last forever, or at least for a very, very, very long time. It's different now, I'm realizing. For generations alive today and those that follow, if we're paying attention, we live with the knowledge that Earth has received a terrible diagnosis. In the previous chapter, we tried to reckon with the tragedy of that. In this chapter I want to talk about how there's beauty in it, too.

A decade or so ago, I wrote a book to help people live with serious and life-limiting illness based on my clinical experience, research findings, and personal mindfulness and compassion practice. Flipping through it recently, I was struck by a passage in the introduction where I'm validating what it's like to come face to face with an incurable illness, but it could just as well describe our collective existential crisis in this historical moment:

> Mentally, you may struggle with feeling unwell and also with coming to terms with the knowledge that you likely will not overcome and be cured of the illness. Your mind may be filled

with fear of the unknown and the future, or with regret for things you have or haven't done. You may feel so anxious about running out of time that the anxiety is paralyzing. You may also feel saddened or depressed by all of the changes and losses before you. Spiritually, you may try to make sense of or question what is happening to you. You may be drawn to explore the meaning of your life and to wonder about your death. Socially, you may notice that your relationships with loved ones and friends are shifting, perhaps becoming more distant or becoming closer. You may realize that you need to figure out new ways to have fun together, be intimate, and have difficult conversations.[15]

Throughout my clinical experiences as a nurse caring for cancer patients and later as a researcher, patients frequently reflected to me that they treasured life more and were living more fully than before their diagnosis. It was an ongoing source of wonderment for me, how even when they were physically unwell and knew their time was limited, they found meaning and joy—and this was true for nearly every one of my patients, not some unusual handful that dealt with their situation exceptionally well. In fact, because it was such a common experience, no single story stands out in my memory, but I have transcripts of qualitative data from more than ten years as a clinical scientist doing research with people with advanced-stage cancer. Thinking about the parallels between people facing their mortality and us, as a species, living with the awareness of so much suffering, death, and possibly our own extinction, I decided to revisit those transcripts.

I studied women with metastatic breast cancer who participated in an expressive writing program, for example, and men and women going through stem cell/bone marrow transplants for lymphoma and leukemia who learned mindfulness meditation. Let me share with you some examples of wonderment that comes from knowing wholeheartedly that life is precious, fragile, and finite, in their words. Each passage is from a different person:

But now, you are made to understand that your days aren't endless. So from that moment on, your outlook on life is forever changed. In a sense, it's a gift because you live your life prioritizing what is important.

Every night I thank God for all he has given me and [for] making me look at life differently. I used to get frustrated waiting in line, traffic, etc. and now these things really do not bother me. I cannot understand, now, why so many people get upset for reasons like the above. I now plant flowers and see the beauty.

It is an absolutely beautiful day, high clouds, sparkling green blowing in the wind. We are all fortunate to be alive and that is the lesson I hope my kids can learn, to appreciate this life and the beauty around them, the joy in the smile of a friend, the sight of the jump of a toad or the blossom of an apple tree.

I appreciate every little thing they (my children) do. There is a heightened awareness to all the small little nuances in day-to-day life. I stay more focused on what's important and what really doesn't matter. Perhaps this is a gift I can pass on to my children.

So, I have learnt much about life due to cancer. I have learnt what is real friendship. I can show my affection for others, people who care.

I try to stop every day and find something beautiful. A gorgeous Japanese maple in an unlikely spot. My son's beautiful smile.

Having cancer has in some ways made me a better person—more spiritual, more appreciative of each day, more attentive to little things around me. But there are days when I wish most fervently that cancer was not part of my life. Then I think of other people coping daily with life-threatening diseases. I am not alone.

So, my blessings for today are: a good support group of friends, a family (and cat) who love me, a beautiful fall day of sunshine outside, and a feeling of well-being—each is a gift.

I love to get away to new places and explore the natural side—the woods, fields, water, animals. It just sets my soul free. Wide-open sky at night, lying in my hammock watching the stars and bats come out.

The other day I was standing at the radiator, and I was just ready to pick something up to read, and I thought, "Ahh, I am at the radiator. It is warm. I am in my home. Isn't this lovely."

We don't—have much time. We have to use it, taste it, feel it, smell it, see it, hear it. It's all we have and all we can do.

There's an oft-told story about a strawberry, that I like. It's an ancient Buddhist parable that you can find in different versions, but in case you haven't heard it, here's mine.

In a faraway land, a woman was trekking along by herself when she noticed out of the corner of her eye that a tiger was following her. She picked up the pace. So did the tiger. As she was thinking about what to do, pretty sure she'd heard somewhere that a human cannot outrun a tiger, she came to a cliff with the tiger not far behind her. However, she noticed a thick vine growing from the top of the cliff, grabbed onto it, and eased herself over the cliff edge just in time. She was carefully lowering herself down the vine when she spied another tiger below her, pacing back and forth and looking just as hungry and scary as the tiger pacing and eyeing her from above.

Between looking down and looking up, the latter was somewhat preferable, but as she did, she saw that a mouse had begun chewing on the vine above her. Now she wasn't just stuck but definitely going to die. Overcome with understandable panic, she took a deep breath, closing her eyes. When she opened them again there was a wild strawberry growing from the cliffside in front of her that she hadn't noticed before. It was right there, big and perfectly ripened. The woman picked what turned out to be the sweetest, juiciest strawberry she'd ever tasted, which she proceeded to eat slowly, savoring each bite.

What tigers are we stuck between as a mouse chews through our time? We think about the past—shame, regrets, kicking ourselves—and think about the future—planning, worries, fear. But all the while, as long as we breathe, the present moment is right in front of us, big and juicy as the incredible fact that we're alive. It's human nature not to notice because we're so carried away by our tigers/thoughts, but with awareness and practice we can spend more time in the present. We find there's nothing like knowing our time is finite to sharpen awareness of what's right in front of us and what really matters.

I remember, in my mother's final days, lying on the bed curled up beside her, my arms gently wrapped around her fragile cancer-ridden body. I softly stroked her skin. I knew it would be the last time she would be with me in that body, a body I loved. A body I came from. This physical connection, this touch and smell, this looking into her eyes and hearing her voice would all be gone soon. Even amid profound sadness and grief, knowing there was nothing I could do to take away all of her pain or stop her from dying, it was one of those moments when, as the poet and environmental activist Wendell Berry says, "I rest in the grace of the world, and am free."[16] Death made the moment precious, one of the most cherished moments of my life.

Every day more species are dying—never coming back. Old trees burning or being cut down. Animals, people included, dying of thirst or smoke-filled air or pollution or flooding. Ice is melting. Waters are warming, rising. I frequently feel like there's nothing I can do to protect them, save them, or stop them from suffering and dying before their time. There's grief deep in my bones. Yet, like in Wendell's poem, if I go outside and I'm quiet, I find "the peace of wild things" is still there. I look around and above me, I close my eyes and listen, and I sense into the life before me, all around me. I breathe it in. I look closely and see the natural kaleidoscope in the center of a flower. I close my eyes and smell it. A bird sings and I listen. Or I take in the sliver of the moon on a clear night, delighted by fireflies. In such moments I feel whole, in love with the world and not needing it to be any other way than what it is.

WHAT A WONDERFUL, WONDERFUL WORLD THIS IS AND COULD BE

Remembering what Donella Meadows, the systems theorist who was also my cancer patient thirty years ago, said about environmentalists failing to project an appealing vision, it strikes me that not all wonderment is concerned with the present tense. As a result of this failure, many people, I think, in specific or vague anticipation of losing what they love about life, resist taking action. They dread what they think that would mean since, as Donella says, they mostly associate environmentalism with restriction, prohibition, regulation. Consider the old stereotype: Environmentalists are smelly and dirty because they don't wash themselves or their clothes often enough, and when they do, they probably use some kind of environmentally friendly "soap" that doesn't actually work. They eat food that isn't very tasty, often brown and high in fiber, and wear scratchy clothes that are unstylish because, as they will be the first to tell you, they're thinking about more important things. They care more about obscure animals than people and are generally a bummer to be around. And if they had their way, their policies would take away our pleasures and freedoms. As Donella says, "Hardly anyone envisions a sustainable world as one that would be wonderful to live in."[17] The alternative Donella proposes is *envisioning*, which in her hands is a kind of wonderment.

First the bad news. In a brilliant talk she gave on the subject (for the full infectious effect, I recommend that you watch the whole thing[18]), Donella says the enemy of envisioning is cynicism, a feedback loop unto itself. And the consequences, she says, "of our culture of cynicism are tragic." Afraid of disappointment or ridicule, we aim low, for what we think we can get instead of what we really, truly want. Then when we meet these hobbled goals, cynicism assumes it was an accident. Whereas when we fall short, cynicism takes it as an omen. So we think we can get even less, and we try even less, Donella explains. "The less we achieve, the less we try. Without vision, says the Bible, the people perish."

"Children, before they are squashed by cynicism," Donella says, "are natural visionaries. They can tell you clearly and firmly what the world should be like. There should be no war, no pollution, no cruelty, no starving children. There should be music, fun, beauty, and lots and lots of nature. People should be trustworthy and grown-ups should not work so hard. It's fine to have nice things, but it's even more important to have love. As they grow up, children learn that these visions are 'childish' and stop saying them out loud. But inside all of us, if we haven't been too badly bruised by the world, there are glorious visions."[19]

It's what we all want, if we're being honest, when we wake up from the cynical status quo and set aside our power trips, isn't it? "I have noticed, going around the world," Donella says, "that in different disciplines, languages, nations, and cultures, our information may differ, our models disagree, our preferred modes of implementation are widely diverse, but our visions, when we are willing to admit them, are astonishingly alike." Greta is officially a grown-up now—she and the Dalai Lama met shortly after her eighteenth birthday, and when this book comes out she will be twenty—but as a child she spoke for not only her peers and future generations of children and grandchildren but also the uncynical children we all were once and maybe still are deep down.

Rebecca Solnit says recently she has seen signs that the environmental movement's message, if not the whole of Western culture, has shifted. "I saw a little 'What are you willing to give up for the climate' kind of questionnaire thing the other day," she recalls, that somebody reversed to ask, "What are you willing to gain to address climate change?" What are we willing to do for politics uncorrupted by the fossil-fuel industry? For air that doesn't kill us and is delicious to breathe? For the ecosystems that we depend on to flourish? What are we willing to do for, as Rebecca puts it, "energy systems that distribute power literally and politically more evenly"? "Maybe," she muses, envisioning is a superpower and maybe "we gain hope in the future."[20]

As the altruistic-minded developer Jonathan Rose likes to ask, "Are you giving up driving or are you giving up being stuck in traffic?" Or as Stephanie Tade knows from giving up alcohol, there was so much more to be gained.

"We should say immediately," Donella qualifies, "for the sake of the skeptics, that we do not believe vision makes anything happen. Vision without action is useless. But action without vision is directionless and feeble. Vision is absolutely necessary to guide and motivate. More than that, vision, when widely shared and firmly kept in sight, does *bring into being new systems.*" [21]

Together with other systems theorists, she says "we mean that literally. Within the limits of space, time, materials, and energy, visionary human intentions can bring forth not only new information, new feedback loops, new behavior, new knowledge, and new technology but also new institutions, new physical structures, and new powers within human beings. [22]

In 2004, Wangari Maathai became the first environmentalist to receive the Nobel Peace Prize. Hearkening back to the stream she played in when she was growing up, the one by the sacred tree, she made the following statement in her acceptance speech: "The Norwegian Nobel Committee has placed the critical issue of the environment and its linkage to democracy and peace before the world. The challenge as I stand here today is to restore this home for the tadpoles and give back to the children a world of beauty and wonder."

We are all Earth's children, even the cynics who mock sentiments like this. Because it's more than a sentiment; it's a fact. What kind of world do we want to live in? What are we willing to gain? What makes life worth living—and saving? What vision will inspire and guide our action?

PART FOUR

Action

So the situation is very serious. In the past, when we humans first began to exploit natural resources in a large scale due to industrial revolution, the impact of human behavior remained less visible. Today, with increasing human population and widespread consumerist culture, this chain of causation has become much more visible. Furthermore, the principle of interdependence means that once a certain causal chain is set in motion, unless mitigating circumstances are introduced, some kind of vicious cycle is created with a momentum of its own. So we must think seriously about the new situation and our lifestyle, and change our way of thinking and living. We must listen to young leaders, like you, for the sakes of our future generations and for the sake of our planet.

—The Dalai Lama

We need to create a social movement. We need to shift the social norms, because if we are enough people who demand change, and who are advocating for these things, then we reach a critical mass and we will no longer be possible to ignore. So that's what we need to do right now.

—Greta Thunberg

7

THE BEGINNING OF
THE AGE OF ENOUGH

What to Do and How to Think About Doing It

"THERE WAS THIS HUGE MOTHER leatherback turtle who had come ashore to lay and bury her eggs," says my teacher and friend Willa Blythe Baker. Willa, the Buddhist lama, is talking about being in Costa Rica recently, in one of its many protected national parks, on an almost-deserted beach. As she arrived, the turtle was on her way back to the sea, but Willa found someone who had witnessed the whole thing and shared with her the video they had taken. "So, I did not know this at the time, but the Pacific leatherback turtle is critically endangered," Willa tells us, "and I was seeing something rare. This was her great return." A pregnant leatherback returns to the same beach where she was born fifteen or twenty-five years earlier, to lay her eggs in the sand and bury them with her flippers. "She knows to do this somehow," Willa says. "It's encoded in her DNA. And she knows that it's time to leave after she's buried her eggs; she goes straight back down to the sea, and she does not come back. And the sun and the sand nurture her babies." When they hatch, they make their own way to the sea.

Willa took this as a teaching: "That mother turtle, though

she's so devoted—devoted enough to come all the way back to her birthplace—she never sees her own young. She will not track their future. But she knows exactly what to do in the moment, and she does the right thing exactly as it needs to be done to ensure the best chances for her babies." Willa says, like the mother leatherback, our job now (and always) is to stay in the present knowing there will be a future but not knowing what it will be— stay in the present and do the right thing.[1]

Listening to Willa, I think about how, unlike turtles, we humans spend a lot of time wondering what to do. I suppose that's especially true for those of us who aren't struggling too hard to meet our basic needs, as well as those of us who have departed from traditional mores in this modern era of individualization and secularization—we feel like we have to figure everything out on our own. What is the right thing? Who knows? Learning more about the climate crisis and naturally being inspired to *do something* can leave us feeling more rather than less confused as to *what*. So finally, let's talk about action. In the book they wrote together, *Our House Is on Fire*, Greta and her family say, "Without action, hope sooner or later comes to an end."[2] I'm not here to tell you to be vegan or never fly in an airplane again (but also, let's not rule out flying less or quitting meat just yet). We don't know where curiosity about doing something will take us. So, let's talk about what to do, what is the right thing in this moment. Let's realize the power and joy of coming together and doing better, since hope and life on Earth depend on it.

When I first told a friend of mine about writing this book and said that I reckoned I would have to be open to letting it change me, she protested, "We can't all be Greta Thunberg." Putting aside the potential for copping out with such a response, there is truth to it. We can't all be Greta Thunberg. Well, then, what *can* we do? Who *can* we be? There isn't one path of right action or one "solution" to climate change. There is each of us and all of us doing what we can do, doing what we're good at, doing what we believe—really, truly, honestly believe—makes sense. As Greta says, we have to do everything we can.[3] And, I would add, we have to ask ourselves what is stopping us from doing anything. Since behavioral science tells us that having options positively affects our motivation to change, this chapter includes quite a few different to-do lists.[4] A choose-your-own-adventure approach to the climate crisis is not only possible but necessary, since no one person or group of people could possibly do everything that needs to be done.

I admit that I approached this chapter with more than the usual amount of writer's trepidation. It wasn't only due to the pressure of a "big finish" or satisfying end. I worried about sounding facile or, worse, *blithe* about what we can do about the climate crisis. You know, like how using biodegradable straws and refilling your reusable water bottle when whole democracies are in danger and the Earth is burning are ridiculously insufficient; or how talking about choices, as a person who has more than a fair share of them, can come off as callously out of touch. I know that, globally speaking, many people can't afford not to live a safe distance from toxic pollution, many more are being driven from their homes by drought or flooding—and the examples could go on and on. At first glance, changing hearts and minds may appear too mushy to have much bearing on doing something in the actual world. And, of course, talking about talking can get you accused of being all talk and no action. But I think talk *versus* action, hearts and minds *versus* the world, and individual *versus* social or political impact are false dichotomies, and I want to talk about that before we get to to-do listing. Plus, at this point we are well beyond first glance, you and I, so I hope you will bear with me.

My friend Roshi Joan Halifax knows what I mean. "There is a reason why many corporations have heavily promoted a focus on individual behavior like recycling or energy-saver light bulbs," she notes, along with pretty much everyone in the climate movement. "And it's not to say that there's anything wrong with energy-saver light bulbs or recycling, but it's also why these companies support autocratic regime change around the world. They don't want us to see that we need fundamental systems change, serious institutional and economic reforms, and for our governments to enforce checks and balances on those companies who profit from polluting our Earth. And they also know that thriving democracies with awake, principled, compassionate, and active citizens are a threat to them. So, fundamentally, the responsibility lies within the human heart and mind."

Let's talk about what the human heart and mind can do.

RECOGNIZE INTERDEPENDENCE

I read an article by the journalist and author Andrew Marantz profiling Sunrise Movement, an international youth-led organization working for climate action (tagline: "We are the climate revolution"). On the way down to Philadelphia from New York City to spend time with a bunch of Sunrise organizers, Andrew ties himself in mental knots of guilt-laced climate bargaining, starting with his driving. "Normally, I'd take the train, or maybe a bus," he says. "Gaze out the window, sample the sluggish Wi-Fi, spend an hour dozing off—before you know it, you've arrived, without feeling too guilty about your carbon footprint. This time, given the pandemic, I drove. It was a beautiful day, so I cracked the windows, saving fuel by forgoing air-conditioning. But, come to think of it, this created drag, which surely made my gas mileage worse. Then again, my car is a hybrid! Maybe I could offset the trip by planting a tree?"[5] Later, the Sunrisers seem to release him from his dilemma and their own; as one of them considers some meat options on a takeout menu that threaten to get him caught up in similar ambivalent carbon calculations, others jump in like a chorus with "a movement adage, using the singsong rhythm of a call-and-response":

The biggest driver of emissions is . . .
The political power of the fossil-fuel industry, not individual behavior!

The reporter's bargaining will sound familiar to anyone who has begun to ask, "What can I do?" Certainly the activists are not wrong to focus their protest on politics and the fossil-fuel industry and, perhaps, to worry less about the occasional beef burrito that comes with plastic cutlery. However, from the Buddhist perspective of interdependence, the whole individual-versus-systems debate, in its tendency to default to the extremes, often misses the truth at the center of things; systems are made of individuals, and individuals are conditioned by systems. The flow of influence goes both ways.

Karen O'Brien and Christine Wamsler are behavioral scientists who want to help us understand the implications of interdependence in secular terms. Their work explores connections between personal and societal transformation, with sustainability in mind. Talking about a shift she made in her career after watching the decades of climate science go by and nobody doing much about it, Karen, a professor in the Department of Sociology and Human Geography at the University of Oslo, says her research "has gone from focusing on climate change to the change part of climate change." She believes that behavioral science can offer insights into how industrial growth societies might get unstuck from their unsustainable ways. "Technical problems can be addressed through improved knowledge, know-how, and expertise," Karen says. "And this knowledge, know-how, and expertise tends to be what we're good at, and we focus research funding and a lot of energy on that. But a different type of problem is an adaptive problem, which is about mindsets—our beliefs, our values, our worldviews in paradigms. And adaptive problems certainly have technical dimensions to them, but if we address an adaptive problem as if it were just technical, we're going to fail."

In search of a more holistic approach, Karen has developed a theory of what she calls "quantum social change," where personal and collective transformation—"inner" beliefs and "outer" actions, individual and societal structures, bottom-up and top-down approaches, *I* and *we*—are interdependent and simultaneous, not opposites or separate spheres. Karen says fractals are a better metaphor for scaling social change: "We see them in nature, we can create them in algebra and geometry. The difference is that social fractals, cultural fractals, and human fractals, they all embed values." And these values, when they guide our action, "resonate and replicate across all scales."[6] Values like "courage, generosity, transparency, humility, humor, and empathy," to take an example she mentions—the stated values of a political movement called Alternative UK. These values "are not just there to be brought out on special occasions," Alternative UK stresses. Rather, they bring a particular quality of agency to the everyday "in the way we think, speak, and act," that scales up through language, behavior, strategies, and relationships from, say, a single email thread to an entire activist campaign.[7]

The point is that who we are and how we show up—every day, each moment, here and now, to everything—matters. "If you think about yourself as a quantum fractal of change that is entangled through language, through meaning, through consciousness, through a shared context, which we are all inhabiting on this planet," says Karen, "then you see that you actually can influence your family and your friends, your personal network, your community, all the way up to the global level." In short, Karen says, "you matter more than you think" (the title of the book she wrote about this), because "the future is conditional, and we are individually and collectively co-creating it through our ideas, beliefs, and actions."[8]

Christine Wamsler agrees that "systemic change and personal transformation are deeply interconnected." Christine is a professor of sustainability science at Lund University in Sweden, and her work also collapses traditional separations between "inner" and "outer" dimensions of change. She, too, sees potential in "addressing mindsets"—that is, "beliefs, values, worldviews, and associated inequalities" that sustainability research has too long neglected. "We know today that climate change and other sustainability challenges are in fact internal human crises," says Christine, "the other side of the human story of separation or disconnection, which assumes that we are all separate from each other, that some humans are superior to others, and that we are separated and superior to the rest of the natural world." Given this, climate action must include changes in consciousness, "in the way we see and relate to ourselves and the world around us, others, nature, and future generations." As the activist Wangari Maathai says, "You raise your consciousness to a level where you feel that you must do the right thing because it is the only right thing to do."[9]

Christine shows me the so-called iceberg model of systems thinking. "The part of the iceberg that is visible above the surface of the sea are the events or crises that define our world today. But much more, around 90 percent or so of the iceberg is invisible to us. So, in human terms, what is hidden beneath the surface are the underlying behavior patterns, social structures, and mindsets that are responsible for creating the events that define our world." Based on this model, says Christine, "we can see that the capacity to reflect on our own mindset and potentially adopt a new paradigm is one of the most powerful ways to dramatically influence sustainability outcomes." She says the iceberg model implies both that we can work toward sustainability at any of the different levels—events, behavior, systems, or mindsets—and that it would be a mistake to overlook any one of them altogether. "If we choose one of these four areas to the exclusion of the others, we might not see the kind of change we're hoping for."

An ancient Jataka tale about a parrot, made popular in modern times by the Zen teacher and author Rafe Martin, has special resonance as a fable in the age of climate change. This is my adaptation of Rafe's telling:

> A brave little parrot lived happily in a forest until one day the forest burst into flames. Being a bird, she was able to fly away to safety on the other side of a river, but as she flew, she looked down and saw that many animals were trapped by the fire and would not be able to escape. What could she do, she wondered, not as a rhetorical question. What she did was dip herself in the river and fly back to shake a few drops of water from her feathers onto the now-leaping flames. Over and over again.
>
> Some gods floating above in their cloud palaces noticed what the parrot was doing and pointed and laughed at the apparent futility of her effort. But one of the gods was moved by the parrot's bravery, determination, and—let's face it—pathos, so moved that she began to weep. She wept down from her cloud and her god tears put out the fire.

What I take from this story is not that the parrot saved the day alone or that there's no need to worry because some greater power will save us, but that the two things go together. I found a modern-day analogy that doesn't involve gods in *Our House Is on Fire*. By the same logic as the people who say what we do as individuals doesn't matter, Greta and her family write, "We could all refrain from paying taxes because 'My little contribution is so ridiculously small in the big picture that it's better that I don't bother with it and invest in things that really benefit me and my family instead. Anything else is virtue signaling.'"[10] Or take flying, for instance, which one by one Greta, her mother, her father, and her sister all committed to not doing. "Staying on the ground creates ripples in the water. And making ripples in the water is the best thing we can do," Greta's mother says. "A friend asks me which flights are unnecessary. My flights, I answer. Just as unnecessary as my shopping and my meat-eating. And no, no one maintains that it's going to be enough. No one believes that consumer power is the solution. But if my microscopic contribution can in some way hasten a radical climate policy, then I'm in."[11]

Rebecca Solnit recalls sitting on the floor of an event with the climate journalist Bill McKibben at the Paris Climate Treaty Conference in 2015 when the umpteenth person approached Bill to ask, "What can I do as an individual?" His answer: "Stop being an individual." Rebecca adds, "Find your people."[12]

DEKILA CHUNGYALPA IS A DEAR FRIEND and colleague and a professional community organizer on behalf of the environment. With a background in forestry and environmental studies, her career has taken shape around the revelation that faith communities have tremendous, too often untapped, potential to effect change. They have already found their people, people with leaders in a position to communicate how climate change relates to them. A Tibetan Buddhist born in northern India, Dekila came to the United States for college and stayed. She now works with faith leaders of all religions to mobilize their congregations and broader networks to work for environmental justice.

At one point she was called back to northern India to train Buddhist monastics in the scientific "underpinnings," as she puts it, of climate and environmental issues. "This is what twelve years of working with faith leaders has taught me, is never underestimate them. I really thought there would be resistance to it. Or that the science might be 'above their heads' or 'above their understanding,' right? Oh my God. Not only did they get it, they started to teach me." Dekila describes moments in those trainings when "the conversation would flip and suddenly one khenpo or one lama would get up and just be like, 'Let me teach you about interdependence.'" She says the trainings had "this amazing energy," and what she had thought was a one-off event took off. "The monks and nuns started initiating projects. And now we have this eco-monastic movement called Khoryug, which has over fifty monasteries and nunneries across the Himalayas, a little bit in Tibet too, doing environmental projects." Projects like reforestation of indigenous trees, rainwater harvesting for safe drinking water, solar kitchens, disaster preparedness, turning monastery flower gardens into organic farms, and water restoration.

We can see the last in that list as one example of the power of quantum change. "I want to take a moment and just situate where the Tibetan community and the Himalayan community is," Dekila says. She is talking about the Tibetan Plateau, the source of many large rivers in Asia. "I'm going to go west to east. From the west, we have Indus, we have Ganges, we have Brahmaputra, we have Irrawaddy, Salween, Mekong, Yangtze, and Yellow Rivers. All of these rivers come from the Tibetan Plateau. Everything that happens in the Tibetan Plateau affects the populations in these river basins, and that is over a billion people that depend on the waters that come from the Tibetan Plateau." So, when a relatively tiny population of monks and nuns work on water restoration at the source of these rivers, it matters. And, as Dekila points out, the effects of their actions multiply not only through the geographical reach of flowing water but through the cultural influence of these particular people as community leaders. The examples set by monks and nuns in these Buddhist communities influence decision-making from the household all the way to the policy level. Dekila says she sees this effect often in her work with faith leaders, and it's a big part of why she finds it so rewarding.

On paper, we can divide societies with lines between the corporations and the people, or between "the 1 percent" and the rest, but on the ground these lines don't exist. The advertising executive and their assistant work in the same place. It may be correct to say that the executive—with their big house outside the city and second home upstate, their daily commute in an SUV, frequent business travel, fancy vacations, etcetera—is contributing significantly to the climate crisis, whereas the assistant who can't afford any of those things isn't, in terms of their respective measurable carbon footprints, but they participate in the same culture. Corporations are made of people. Chances are the assistant aspires to live like the executive. This is not to say that class differences and power differentials aren't real, but I don't see how the norms, values, and standards shaping the behavior of corporations or the 1 percent could change without pressure from everybody, pressure that comes not just from awareness but from lived, embodied, example. Unless we all prize very different things, how else will the 1 percent get the message that money, power, growth, fame, and luxury consumption do not mean they are winning at life? That, in fact, by not participating in the culture of enough, they are missing out on richer sources of connection and meaning?

Greta's dad, Svante Thunberg, is driving home from London.

Late in the evening, queuing for coffee at a McDonald's at the Hamburg Süd truck stop, Svante tells a man in broken German that he is on his way from London to Stockholm in an electric car because he has stopped flying für das Klima, and although the man understands what he's saying, he doesn't understand what he's saying, and there in the car park in the wind and the rain, Svante cries openly for the second time in fifteen years. Because there, surrounded by 50 billion lorries, motorways and BMWs, he realizes that it doesn't matter how many electric cars we acquire. It doesn't matter how many solar panels we put up on the roof. It doesn't matter how much we encourage and inspire each other. And it doesn't matter if we stay on the ground and renounce the privilege of flying, because what's needed is a revolution. The greatest in human history. And it has to come now. But it's nowhere in sight. For five minutes he stands there, until he realizes that no one can live with the thought of giving up. And that nothing will be solved by crying at German petrol stations. All there is to do is drive on. Towards Jutland. Towards Malmö. Towards the dawn.[13]

When I feel small and at the mercy of a merciless system, I will try to remember what Karen, Christine, Bill, Rebecca, Dekila, and Greta and her family are saying. Ripples or fractals of change—and gods if you believe in them and act on your beliefs—are big enough to make a difference, and these ripples come from me and you, though we can't say exactly where they're going. We don't know what will happen when we open ourselves to caring and learning about climate change. We don't know whom we'll meet when we start taking our kitchen scraps to the community compost bins at the farmer's market. We can't be sure of the effect it will have to write about the climate crisis in our church newsletter or show up for a protest march or a city meeting. We don't know who is going to win the election when we vote. I don't know if anyone is going to read this book. We don't know how this conversation is going to go. We don't know what will happen when we "confront the bus driver," as Wangari urges us to do when we feel like we're on a bus going the wrong way. We're going to have to trust.

When I feel small, I can reflect on the fact that I'm one of eight billion people on the planet. This geological age we are in is being called the Anthropocene in recognition of the fact that for the first time in history, humans are the dominant influence on the environment and climate. Isn't it ironic, then, that we in the Anthropocene would think we can't make a difference? As much good as I can do times eight billion sounds pretty good. Instead of "It won't make a difference if I do this one little thing," I can do one good thing and then ask, "What else can I do"?

When I feel small, I will think about the fact that Greta didn't know what would happen, didn't know that she'd be "big." She was a fifteen-year-old girl with a typical fifteen-year-old girl's Twitter following when she sat down alone in front of the parliament building with her conscience, her handmade sign, and her smartphone. Like the mean gods in the story, many people pointed and laughed, but she was undeterred. She was a brave little parrot. We can be, too.

"We can't solve a crisis situation unless we treat it as a crisis situation," Greta's mom, Malena Ernman, says. "Everyone who has ever witnessed an accident knows what I mean. In a crisis we get superpowers. We lift cars, fight world wars, and climb in and out of burning houses. It only takes someone falling down on the pavement for a line of people to appear, prepared to drop everything to help out. It's the crisis itself that is the solution to the crisis. Because in a crisis we change our habits and our behaviour. In a crisis we are capable of anything."[14]

ONE HUMANITY PRACTICE

His Holiness the Dalai Lama

In my own practice, one of the things I find most helpful is to cultivate an altruistic outlook on a daily basis, as if charging my inner battery. Every morning when I get up, I remind myself that I am just another human being, one among so many fellow humans on this Earth. We are all the same—each and every one of us; we all wish to find happiness and do not want suffering. As social creatures, we seek connection with others and find joy through our relationships. Our well-being is deeply interconnected with the well-being of others. In fact, there is no such thing as "my own interest" independent from others. To help me remember this, I reflect upon certain passages from ancient Indian masters who recognized how, when we neglect this fundamental truth of our interdependence and become excessively self-focused, we sow the seeds of suffering. In contrast, orientation toward others, caring for others' welfare, opens the door to our own well-being. In particular, I recite the following lines from the eighth-century Buddhist teacher Shantideva:

> Whatever suffering there is in the world,
> it all stems from self-centeredness;
> whatever happiness there is in the world,
> it all stems from wishing happiness for others.
>
> What more needs to be said?
> Just look at the difference between the two:
> The childish who seek their own well-being
> and the Buddha who seeks others' well-being.

Therefore, if one does not switch this outlook
of self-centeredness to other-centeredness,
let alone the attainment of buddhahood,
even in this life there will be no real joy.

If we reflect deeply and examine the psychology of self- versus other-orientation, we'll come to recognize that many of the fears, anxieties, and stresses that plague us are rooted in excessive self-focus. Excessive self-focus makes us tense, our sense of self fragile, bringing ego brittleness, and it leads us to overreact in the face of a challenge. On the other hand, if we can open our hearts a little bit and orient toward others, this simple shift makes us feel more expansive, even more courageous, I would say. There is a sense of freedom that comes in the absence of the heaviness of self-agenda. We can feel more relaxed even. Also, being in touch with our shared humanity with others, we feel less lonely and more connected with others. So, if we are truly serious about our own long-term well-being, we need to open up our heart and orient more toward others. And, of course, if we truly care about the world and its inhabitants, compassion has to be at the core of how we view the world and relate to others.

It is for reasons such as these that in the Buddhist tradition there is a great emphasis placed on cultivating compassion and the altruistic awakening of heart and mind, a practice traditionally called *bodhicitta*, which I do every morning.

This teaching that orientation toward others' well-being is the key to greater joy is something I find deeply meaningful and inspiring, and powerfully true in my own personal life as well.

REALIZE WE ARE PART OF NATURE

Dekila lives in Wisconsin now, since home base for her work with faith communities, since she founded the Loka Initiative, is at the University of Wisconsin–Madison. But she grew up in Sikkim, a small state in the far north of India, the tip of a baby finger of land that sticks up between Bhutan to the east and Nepal to the west and touches Tibet. "When I was born," she says, "Sikkim was still an independent kingdom." She talks about how she grew up in a strongly Tibetan Buddhist family—her mother became a nun when she was eight or nine, as her grandmother had also done. "Our community is Bhutia, which means 'from Tibet,' and is one of the three Indigenous communities in Sikkim."

The third-tallest mountain in the world is in Sikkim, but maybe you haven't heard of it. Kangchenjunga is not nearly as famous as Everest or K2, and Dekila says there's a reason for this. "The Indigenous people of Sikkim have held steadfast against any kind of mountain-climbing expeditions, despite all kinds of lucrative offers that you can imagine. We have refused to allow it to be climbed." Why not? Because, she says, "our identities are infused with the land, and we see nature as having sacred significant living value." Like the wild-rice-growing communities around the Great Lakes that Kyle Whyte talked about and the "tree of God" that Wangari Maathai grew up with in rural Kenya, the Indigenous people of Sikkim understand their relationship with Kangchenjunga, and nature in general, as one of interdependence and kinship. This, Dekila says, "results in protection of biodiversity and protection of nature, protection of all other nonhuman living species. For me, this concept of interdependence was something that I carried all day long. The way my mother taught me was to always remind me that even the oxygen in my lungs came from outside and therefore life was a gift; it was not something that I had any rights to, it was something I needed to feel gratitude for.

This deep love and reverence for nature was deliberately inculcated in us children when we were growing up." There's a reason that though the world's 370 million Indigenous peoples make up less than 5 percent of the global human population, they manage and protect 80 percent of the global biodiversity, as Lyla June also noted.[15]

Like individuals versus society, humans versus nature is a false dichotomy with very high stakes. The natural world invites our partnership each time we breathe in. Earth wants to work with us; and from my conversations with Dekila, Kyle, Wangari, Lyla June, Vandana Shiva, and many others, it is clear to me that one thing we can do is take Earth up on her invitation, listen to what she is saying to us, and realize we are part of nature, no matter where we live. Our separation—or rather our imagined separation—from nature makes Industrial Growth Society possible; an adaptive problem that cannot be solved by technology alone since the solution to an adaptive problem depends, as Karen says, on our values and beliefs. "I often joke that the moon and stars look beautiful," adds the Dalai Lama, "but if any of us tried to live on them we would be miserable. This blue planet of ours is a delightful habitat. Its life is our life; its future our future."[16] What would the future look like if we really believed this? What if we carried this awareness all day long and taught it to our kids?

We don't have to camp in the wilderness or travel to a national park to realize we're part of nature—that's the sense of separation talking, when we feel like nature isn't already here, right where we are. Breathing is one way to experience ourselves in relationship with the elements, and it doesn't require formal meditation to simply appreciate the dissolving of boundaries into a reciprocal lifeline that happens with the intimate give-and-take of breath. If you catch yourself taking breath for granted, try holding it for thirty seconds! Or look at a green plant and think about the incredible fact that you are breathing together, like a dance.

Lyla June says eating is another way to connect, when we bring a certain quality of awareness to it. "The word *food*," she points out, "is a noun [in English]—it's an object, it's static, it's lifeless, it's dead. But in our languages, food is always a verb, in the sense that it's derived from a verb. It's a noun that's created from a verb. This is because for us, food is a dynamic, living process that is constantly in flux. As a Native person, you are not just thinking about the nut you're eating; you're thinking about the ancestors who planted that chestnut tree sixty years ago and did so with ceremony and with song.

"You're thinking about how you burned around the chestnut tree to prevent overpopulation of the forest and to return nutrients to the soil and to smudge the trees—it's a ceremony when you burn around the trees.

"When you look at that nut, you're thinking about the rains that came, and you're thinking about the mycelium that nourishes and sustains the soil for the chestnut tree. You're thinking about going out in the forest by the dozens and harvesting chestnuts as a community. You're thinking about shelling them, you're thinking about grinding them, you're thinking about processing them, you're thinking about mixing them with other foods to create superfood mixtures. You're thinking about the spirit of that chestnut tree and how she's like your mother. You're thinking about how we planted chestnut forests with our bare hands and always spacing them far apart so the disease couldn't travel through them, to keep them healthy. You're thinking about the plants that are sisters and brothers to the chestnut that grow around the tree."[17]

By contrast, when we toss a plastic baggie of Cheerios to a child in the back seat, we're not usually thinking about anything like this, let alone teaching the child about their place in nature. This matters because, as one of the founders of Greenpeace, Rex Weyler, talks about, we are suffering from "ecological trauma"[18] and loneliness as a species. It also matters because only when we separate ourselves from the sources of our food can we remain unbothered by the destructive truth of our factory farming and industrial food systems. As Greta and her family point out, this myth of separation, this delusion of human independence from nature has an "astronomical economic value" in Industrial Growth Society.[19] Powerful interests work to perpetuate the myth, and so waking up to the truth of being a part of nature is a radical act. So, it really does matter if you become willing, and stay willing, to buy organic and to advocate for meaningful enforcement of organic standards, for example. It really does matter if you start to see your local area as a watershed and a food-growing region and get to know what that means and who's already doing good things that you can support.

"Many of us long to connect to nature as if nature exists somewhere else," says Sebene Selassie. "Out there, in the countryside or on a mountaintop. Nature lives at the beach or maybe in the park. Not here on the grimy sidewalk or in my noisy apartment." Nope, she says, "here's the deal: We ARE nature." She elaborates: "I experience the water element when I fill the kettle, fire when I light the stove, air when I open a window to feel the breeze, and earth in the wood floor beneath my feet. When I feel especially airy, I can emphasize the groundedness of earth. When I feel too fiery, I can encourage the moistness of water. Parts of me (my breath, skin cells, heat, and tears) continually mix and meld with the rest of nature. I am nature. I belong to it all. You too. You belong."[20]

Yes, we are nature! As I know well from training in nursing and psychoneuroimmunology, being a part of nature is not a lifestyle choice or cultural bent; it's a fact. Every cell, organ, and tissue, and all the fluids in our bodies are composed of natural elements, and these elements—the air we breathe, water we drink, nutrients in the food we eat, and even sunlight—are essential for our health and survival, for our being. Reflecting on that, I understand that it's not a choice to be part of nature, since we all already are. The only choice is *how*: respectfully, graciously, intelligently, and mutually? Or mindlessly and exploitatively? Sustainably or unsustainably? The Dalai Lama says, "On a close examination, the human mind, the human heart, and the environment are inseparably linked together. In this sense, environmental education helps to generate both the understanding and love we need to create the best opportunity there has ever been for peace and lasting coexistence."[21] As I worried about writing this chapter, I could hear the cynics mocking the idea of "reconnecting with nature" as woo-woo and woefully insufficient. But when the Dalai Lama puts it like that, it doesn't sound so blithe, does it?

GO TO THE EARTH IN THE MORNING

A practice from *Lyla June*

Indigenous musician, scholar, and community organizer of Diné (Navajo), Tsétsêhéstâhese (Cheyenne), and European descent

"Ceremony is how we remember to remember." This is what I was told by my mentor Dr. Gregory Cajete of the Santa Clara Pueblo Indigenous nation. Every single thing we do in Indigenous cultures, generally speaking, is a prayer. We never separate prayer from not-prayer. When the anthropologists first came to my people, they thought we didn't have a religion, but the truth is the ceremony just never stopped. We were always engaged in the ceremony of living, of being.

If you don't have ceremony anymore because the British

army stole it from your ancestors a thousand years ago, or the Romans stole it two thousand years ago, or the system of American slavery stole it from your people four hundred years ago; if you are missing ceremony in your life to regularly remember your connection to this Earth, the Sky, the stars, our fellow human and animal creatures and other spirit beings, one thing you can do is go to the Earth and offer something in the morning. This is what my people do to remind us to remember who we are and what our prayers are each day.

Choose a plant, food, stone, or something else that is precious to you and resonant with nature. Whether we are descendants of Turtle Island (a.k.a. America), Africa, Europe, Asia, Pacific Islands, or Australia, we all have ancestral elements that heal our bodies and help us connect with our ancestors. Scarlet Globemallow, for instance, is indigenous to the Southwest, where my Diné ancestors lived and where we still live. It could be anything.

Before you get your day going, look to the east where the sun rises.

Put your precious thing down in front of you and offer it to the Earth. Say what your intentions or prayers are for that day. Invite your ancestors to be with you. Invite the good spirits to walk with you, the spirits that want to help.

Honor who you are and honor the Earth and honor the spirits and the great Creator—or life forces or unknowable mysteries or God or gods or whatever it is that you believe makes the world more than meets the eye.

I think of this practice, and ceremony in general, as a kind of string that holds the beads of our life together. If you can do it every day, great, but even once a week or once a month will help you remember. I pray it helps you remember to give your best but also to receive—the sunlight, the birds, the fruit of joy, the fruit of life, the answers to your prayers, whatever they may be.

FIND OUR WAY INTO THE CONVERSATION

"It is with acts of attention that we decide who to hear, who to see, and who in our world has agency," says Jenny Odell. "In this way, attention forms the ground not just for love but for ethics."[22] And I would add, for action. Since conversation gathers our attention, joining the conversation about the climate crisis is something we can do about the climate crisis. In a review of Jonathan Safran Foer's book *We Are the Weather: Saving the Planet Begins at Breakfast*, the climate journalist Kate Aronoff pronounces, "If the world does manage to steer away from catastrophe, the credit will be owed to a critical mass of social movements, unions, and the elected officials accountable to them, working to take power back. No angst-filled breakfast or lunch can do the same."[23] I take her point that all talk and no action won't solve the climate crisis, but I don't see them in opposition; talk may not be sufficient, but it is necessary for action. People who are doing something about climate change are also talking about it, and I'd venture the converse is largely true, too—that people who aren't talking about it aren't doing much about it either.

The public health scientist Edward Maibach's work explores ways to get more people talking and doing something about climate change, and he has gathered enough data to hypothesize five key mechanisms for building public will. First, he says, "we can increase the number of people who see their community as being harmed already." Second, "increase the number of people who will understand that attractive solutions already exist in a theoretical context." Third, "increase the number of people who feel hopeful as opposed to hopeless." Fourth, says Ed, "increase the number of people who actually publicly express their concerns and their hopes. I'm calling that disclosure, I'm not sure it's the best term but we know that people who are talking about this issue, their concerns and their hopes, they are not only more likely to take action; they're more likely to *convince their skeptical friends and family members* that this is a problem worthy of solution." (Emphasis mine.) "And finally," says Ed, "we need to increase that proportion of people who feel that acting together, this is a problem that we can address. A sense of collective efficacy." In fact, all five of these ways to get people to engage depend on communication.

There isn't only one way into the climate crisis conversation, I've discovered. Thank goodness, because I used to think it was all about reading deadly serious news articles and IPCC reports or making myself watch documentaries I dreaded seeing. The headlines alone, just the thought of them, made me want to run away and hide. Then I realized that the best way into the conversation is the way you will actually take. So, I'd like to talk about some approaches I've come across and some ideas from my personal experience finding my way in.

Educate Ourselves

It's necessary to educate yourself, but it is not necessary to have all the facts before you get here. Do not confuse these things. We don't all need to be experts. You will learn as you go, as you get more and more interested. On the other hand, you won't know what ways into the conversation exist without learning a little first. This book is meant to provide a little learning as an entry point. If you've read this far, you've found a way in! And you can keep reading for more suggestions, to-do lists, and resources to keep going. But not everyone will be inspired by a conversation between Greta Thunberg and the Dalai Lama. To some people, for example, this particular way in may seem "too Buddhist." Someone I know, a practical-minded professional dog trainer, asked me with a slightly wrinkled nose what spirituality has to do with it. There are other books. The trick, I discovered, is not to force it. If a title puts you off, if *The Sixth Extinction* or

The Uninhabitable Earth is too scary for a starting place (though these are both excellent books), find something different. There are books like *Regeneration* or *All We Can Save*, equally excellent on climate change, whose titles are less likely to ruin your mood before you've picked them up. Education needn't be bad-tasting medicine.

Someone else might feel daunted by the prospect of reading a whole book to begin with. That's okay, since there are so many other options in length, format, angle, tone, and so on that, with a bit of poking around, I imagine anyone can find an entry point that feels inviting to them. If you know someone who doesn't want to read this book, when you finish it you will know enough to point them to other interlocutors and sources. Podcasts have been a revelation for me. These days, I no longer hide from every climate news article, since joining the conversation has made me less afraid. It has led me to people who have shared their coping skills and compassion, and now I know I don't have to be alone with whatever information I find. But there is something uniquely approachable about listening to other people speak about this subject with each other, not as a formal presentation directed at an audience but as a connection between two human beings. Podcasts give you instant access to the conversation without any pressure on you to say the "smart" or "right" thing. And since you can do other things while you listen, like cook or garden or fold laundry, with podcasts you don't have to be completely committed to taking time out of your precious and busy life to learn about climate change.

In retrospect, I realize that being too afraid to learn anything about climate change or talk to anyone about it was its own feedback loop of ignorance, fear, and isolation. But the loop can go the other way, individually and collectively. The more people who join this conversation, the more ways we'll find to talk about it and the more people will feel, if not comfortable exactly, like they can be in it and be themselves. The more people who join this conversation, the broader, more inclusive, and more diverse the definition of "environmentalist" will become. The more people who join this conversation, the more registers we'll have through which to connect with one another and express ourselves. I mean, 2021 brought us the movie *Don't Look Up*, a feature-length *comedy* about climate change (that is funny but not at all blithe). The more of us there are, the more people will find a way in.

My dear friend and colleague Barbara (Bobbi) Patterson is a professor emeritus of pedagogy at Emory University. Her work focuses on effective teaching and learning at the undergraduate and graduate levels, as well as contemplative teaching and learning strategies including ethical decision-making. Bobbi says learning about places, starting where we are, is a great way in. "Places invite our presence," she says invitingly. "They call us to just actions when we're asked to engage with the structures and power of place." We can reflect on the history and ecological inheritance of the place where we live—its geological history, its plant and animal histories, and the history of those who have lived there over time. How have they shaped the present reality? I can ask myself, Who am I now in my place? How might I nurture a greater sense of collective belonging and stewardship of the land where I live? What does my sense of connection to the place I live and the stories I learn about it reveal about where healing is needed? According to Bobbi, "Taking places and their stories seriously is fundamental to planetary healing." She suggests forming a place-based group to explore these questions together. This reminds me of Greta and her father traveling to the mountains north of Sweden to learn about the impacts of warming there and to see for themselves—a trip beautifully retold in *Our House Is on Fire*. I think, too, of the Dalai Lama's advocacy for the environmental protection of his homeland, Tibet.

As Christine and Karen talked about, and as this discussion

about educating ourselves has already suggested, facts of the situation "out there" aren't the only kind of knowledge we need for conversation and action. It also helps to know ourselves, to identify the most natural and useful ways we can engage—or more to the point, *will* engage. What are our strengths? What, specifically, can we contribute? What really matters to us? What are our priorities, and how do they relate to climate change? What motivates us? What settings bring out our best? At the same time, is it possible that who we are and what we do are not as fixed as they seem? And critically, what is getting in the way of doing what we already want to do? If guilt, fear, or compartmentalization is paralyzing you, how might you deal with it? What would ease it? If you're compulsively shopping online for things you know you don't necessarily *need*, what's that about? If your business operates in harmful ways, what is stopping it from changing? We can ask ourselves these questions individually and collectively. Behavioral science tells us that people tend to put too much stock in carrots and sticks to overcome resistance—incentives to do something, disincentives to stop doing something, and more and more argumentation about why—when often the problem is better addressed by removing friction or "the obstacles we don't see."[24] For example, is a lack of knowledge about the climate crisis, or disbelief in humans' capacity to change, or insufficient will to join the conversation getting in your way?

I GATHER FROM THE COMMUNICATION EXPERTS in this conversation so far that equally important to knowing ourselves is getting to know the people we're talking to. The Paris Climate Agreement mediator Christiana Figueres says for her, it's all about questioning, not judging. "I don't know why someone continues to eat meat," she reflects. "I just simply don't know. So, my approach is one of honest interest, curiosity, if you will. Tell me what happens for you before you eat the meat, while you're eating the meat, after you eat the meat. Just let me into your reality." If your local action society doesn't want to talk about climate change, find out what they do think would improve their quality of life and connect it to climate solutions, she says—*without using the words "climate change!"* Christiana calls that *contextual communication.* "It's very important to understand what is important to the person you're speaking to. Put yourself in that person's shoes." Or that country's shoes.

"Take Saudi Arabia," she offers by way of example. "That's the only source of its wealth, the export of oil and gas. And it totally depends on that. You can imagine that any idea of starting to reduce oil and gas in our economy is a huge threat to them. But I also understood that this is their current reality. It is not their fault. I cannot hold them responsible for having discovered that resource under their feet. That's the way it happened. All those dinosaurs died there millions of years ago, and they discovered that. So, I don't hold them responsible for that. I hold them responsible for what they're going to do in the future. And so my conversation with them was always about the future, not blaming them for the past but always about the future. And how do they see themselves ten, twenty years from now? Yes, they have made a lot of money out of oil and gas, but is that really going to be the most important thing for them in twenty years? Already, today, they're in one of the hottest areas of the planet. And if we continue to warm, that whole area will be rendered practically uninhabitable. Twenty years from now, where's the water going to come from? Where's the food going to come from?" Christiana says these are things they were willing to talk about. The challenge in getting people into this conversation and keeping it going is not to box them in with guilt but to invite them, as Christiana does, to "paint for me the future that you want to see for your children, your grandchildren, and generations after that. What are you creating for them?"

"Right now," she adds, "we're living different realities, but eventually the future coincides for all of us, because we're all on this tiny, tiny little planet." Eventually the future coincides for all of us—this sounds like a powerful idea to me.

Katharine Hayhoe, the climate scientist and educator we met in chapter 4, agrees. "The most important thing to realize is that everybody already has the reasons they need to care. They might not be the same as ours. And that is okay. But if we are a human living on planet Earth, if we are someone who cares about our family or the safety of our home, if we're somebody who has something that we enjoy doing outside, or we're a member of the Rotary Club, or we're a member of any major world religion, which has as part of its tenets of faith the idea of being good stewards or caretakers of creation and the concept of caring for those less fortunate than us, whoever we are, we already have every reason we need to care about climate change. So, when we talk to people about climate change, it's not about convincing them to care for the same reason we do, because good luck with changing anyone's values once they're past childhood. No, it's much easier than that. It's about figuring out what they already care about, because climate change is affecting everything that's already at the top of their priority list today." Then, she says, it's a matter of connecting the dots between what they care about and how climate change is affecting that and talking about constructive responses that we can make together. Often, Katharine notes, these are things other communities are already doing.

I take this to mean that when Greta says the climate crisis is all we should be talking about, it doesn't have to be in so many words. We can talk about what we really care about. We can talk about how we want to help each other and what we want to do together. It's bound to be relevant since everything that's important to any of us is already about climate change.

Jenny Odell makes one more point about conversation, which is to remind us in this age of likes, follows, and retweets that quality, not just quantity, matters. She offers this example in contrast to the internet (granting that the example might be "over-determined"): "If someone made a zine and they only mailed it to twenty people, but those were twenty people who were going to sit down and spend time with it and talk to other people about it, maybe write something in response—and they respond and then you get somewhere. Like, there's traction. That's a different way of measuring impact. I don't even know what to call it, but that feels substantive to me. And I'm kind of more interested in that and bored with the other thing." The other thing being social media.

Jenny speaks of these different qualities of attention and conversation in terms of feedback loops, actually. In our isolation, loneliness, and dread, we turn to social media for company: "You want to feel some connection; and you want to feel validated, seen, and recognized, but you don't get that." What we get makes us feel *more* anxious, *more* disconnected, and lonelier—this has been shown in studies, she says—so we're even more driven to seek connection. As long as we keep seeking on social media, it's a vicious circle. But we can break the cycle. We can put down the phone and choose something else to give our attention to and turn the cycle in the other direction. "You find other sources of meaning and belonging and being seen, and that makes you feel more stable. And then, because you feel more stable, you don't feel the need to go to social media." As her social media feeds have become much less interesting, she noticed it freed her attention for other activities, like watching birds, writing a book, or organizing.[25]

MAKE A TO-DO LIST

I do not feel qualified or inclined to boil down all the suggestions for action that I've heard and read about into one master list. For one thing, there are many solutions to this problem—there are going to have to be—and no one person can do them all. Then, as we've also talked about, when we have multiple options to choose from, we are more likely to pick one, feel invested in it (because we chose it), and do it. Paul Hawken acknowledges, "Because of the differences among people, cultures, incomes, and knowledge, there is no one common or correct checklist. The top 'ten' solutions to reverse global warming are an abstraction. The true top solutions are what you can, want, and will do."[26] In that spirit, here's a wide sample of to-dos from several different highly qualified people, including Paul, ranging in flavor from voting to finding community to "planting enormous quantities of trees."

Five Things To-Do List

A Canadian study published in 2017 in *Environmental Research Letters* determined the top-five things we can do to reduce our carbon footprint, if you're wondering what changes would make the most difference.[27] To paraphrase, the to-do list suggested by the study is this:

- ▶ Drive less.
- ▶ Fly less.
- ▶ Have fewer children.
- ▶ Eat less meat.
- ▶ Educate teenagers about these things.

"Have fewer children" is controversial.[28] In the context of this study it applies to people in developed countries, each of whose offspring is expected to add an average of 58.6 tonnes of CO_2-equivalent emissions per year to the family footprint. Families in developed countries have fewer children compared to less-developed countries, but their carbon footprint is higher. The author and activist Genevieve Guenther, referring to the study, calls this advice "misanthropic" and "just wrong," arguing that it's about the consumption, not the kids. In a podcast with Genevieve, the climate journalist David Wallace-Wells adds that "if we get to a place where we have decarbonized much of our economy, which is technologically and politically possible now, then you're talking about multiplying invisible carbon footprints."[29] Aw, little invisible baby carbon footprints! But we're not at that place yet. And I disagree that it's misanthropic to consider human impact on the environment as a factor in family planning; from the perspective of interdependence, isn't it shortsighted—narcissistic, even—of our species *not* to think about it? Current carbon footprints in rich countries aside, there are places around the world, global north and south, where the land no longer supports large families, even though their individual carbon footprints are relatively small.[30] The way I see it, nobody should tell anyone else how many children to have, but people everywhere should have the education and means to make informed choices.

Paul Hawken's Regeneration Guidelines and Punch Lists

Paul Hawken says we need to do three things to stop global warming: "First, reduce and eliminate emissions from fossil-fuel combustion. Second, sequester carbon into the soil by photosynthesis in grasslands, forests, farmlands, mangroves, and wetlands. Third, protect the carbon here on earth."[31] How do we do this? In his book *Regeneration*, Paul offers a checklist that we can apply "to every level of activity: people, homes, groups, companies, communities, cities—and countries too." His is a list of questions to guide our every action. "The number one guideline is the fundamental principle of regeneration. The remaining are outcomes of that principle."

1. Does the action create more life or reduce it?
2. Does it heal the future or steal the future?
3. Does it enhance human well-being or diminish it?
4. Does it prevent disease or profit from it?
5. Does it create livelihoods or eliminate them?
6. Does it restore land or degrade it?
7. Does it increase global warming or decrease it?
8. Does it serve human needs or manufacture human wants?
9. Does it reduce poverty or expand it?
10. Does it promote fundamental human rights or deny them?
11. Does it provide workers with dignity or demean them?
12. In short, is the activity extractive or regenerative?

Then Paul recommends that individuals, groups, and institutions make "punch lists" of actions they commit to taking within a certain timeframe—whether a week, month, year, or five years. The Regeneration website has examples of punch lists and a worksheet to help you come up with yours at www.regeneration .org/punchlist.[32]

Ayana Elizabeth Johnson and Katharine K. Wilkinson's To-Do List

This list comes from the anthology *All We Can Save: Truth, Courage, and Solutions for the Climate Crisis*—edited by Ayana Elizabeth Johnson, a marine biologist, environmental strategist, and policy expert; and Katharine K. Wilkinson, a writer, activist, and cofounder of the All We Can Save Project—and the poetic descriptions of the book's part titles the editors offer in their introduction.[33]

BEGIN

1. Root
A call, a welcoming, a place to ground
The foundation of Indigenous wisdom
And the wisdom of Earth's living systems
Interconnection, emergence, justice, regeneration

2. Advocate
Strategy, participation, public good
Plying tools of legislation and litigation
How we hold the powerful to account
And (re)write the rules with all people in mind

3. Reframe
Language and story, creativity and culture
Our means of making sense
To tell the truth—expand, flip, and rekindle it
Imagining, evolving, holding on to our humanity

4. Reshape
Problems embedded in the contours
Of cities, transport, infrastructure, capitalism
Coastlines and landscapes where human-nature meet
Much to reconsider, rend, invert, remake

5. Persist
Damn if this work isn't hard, our task towering
The fire of activism—on the front lines, in the belly
Standing for justice, for health, for the sacred
We don't have to do this alone

6. Feel
Awake, aware, attuned
Hearts break, souls shake with anxiety
Can't skip this: struggling, mourning, raging, healing
A ferocious love for the planet we call home

7. Nourish
Soil, food, water, sky—inseparable
The foundations of our aliveness
Collaborating with and supporting nature
Microbes, farmers, photosynthesis

8. Rise
Generations—growing, giving, gathering
Nurture community and transformation
For a future that holds us, all of us
This is the work of our lifetimes

Onward

Catherine Ingram's To-Do List

Catherine Ingram joined our conversation in chapter 5 to help us talk about the heartbreak of climate change. This is an abridged version of the pointers she offers in her article "Facing Extinction."[34] Despite her grim prognosis, she does not conclude that there is nothing we can do—far from it.

▸ **This is now the time to give yourself over to what you love**, perhaps in new and deeper ways. Your family and friends, your animal friends, the plants around you.

▸ **Find your community (or create one).** People are beginning to wake up and speak about this all over the world. Extinction Rebellion, which began in the UK, has gatherings in many cities of Europe, North America, and Australia. There are also online extinction-aware Facebook groups, such as Near Term Human Extinction Support and Faster Than Previously Expected. You may want to start discussions in your own home with friends and neighbors.

▸ **Find your calm.** In addition to wisely directing your attention, include also whatever daily activities induce greater calm in your life—walking in nature, a slow meal with loved ones or on your own, reading or listening to music, dancing, swimming—whatever your thing is, give priority to it every day.

▸ **Release dark visions of the future, and pace your intake of frightening news.** Although frightening pictures about what is to come in the future may arise in your imagination, it is best not to entertain them. Have a fast from the news as needed and rest your weary mind.

- **Be of service.** Know that whatever is to be in the future, it will feel good to be of service in whatever ways your gifts can be used and on any scale that feels right and true, whether in your personal life of family and friends or in a larger community. There is no need to keep accounts of whether your actions will someday pay off.

- **Be grateful.** Longevity was never a guarantee for anyone at any time of history. Whatever time is left to us, we are the lucky ones. We got to experience life, despite the overwhelming odds of that not being the case. When we think of all the times our ancestors had to thread the needle of survival and live long enough to procreate, *every single lifetime*, it puts into perspective how precious is this experience we are having. Gratitude for life itself becomes the appropriate response.

- **Give up the fight with evolution.** It wins. The story about a human misstep in history, the imaginary point at which we could have taken a different route, is a pointless mental exercise. Our evolution is based on quintillions of earth motions, incremental biological adaptations, survival necessities, and human desires. We are right where we were headed all along.

Donella Meadows's To-Do List

In *Limits to Growth*, Donella Meadows and her colleagues mention "many things to do to bring about a sustainable world," which I've fashioned into a bulleted list. "All people will find their own best role in all this doing," they say encouragingly. "We wouldn't presume to prescribe a specific role for anyone but ourselves. But we would make one suggestion: Whatever you do, do it humbly. Do it not as immutable policy, but as experiment. Use your action, whatever it is, to learn."

- New businesses have to be started and old ones have to be redesigned to reduce their footprint.
- Land has to be restored, parks protected, energy systems transformed, international agreements reached.
- New farming methods have to be worked out.
- Laws have to be passed and others repealed.
- Children have to be taught, and so do adults.
- Films have to be made, music played, books published.
- Websites [have to be] established, people counseled, groups led, subsidies removed, sustainability indicators developed, and prices corrected to portray full costs.[35]

Genevieve Guenther's To-Do List

This one is adapted from Genevieve Guenther, founder of the organization End Climate Silence. "Pick one," she says, "do it once a week, and things will change."[36]

1. **First thing is vote.** You can't do that once a week, but vote in every election. Vote for candidates committed to climate, and then once they're in office, keep them accountable to their promises.
2. **Join a campaign or an activist group.** There are local chapters of groups called the Sunrise Movement and 350 .org in many communities.
3. **If you don't have the time to do that, donate money.** Donate money to organizations that are putting their bodies on the line. Here are some of them: Sunrise Movement, Fridays for Future (the youth organization organizing the climate strikes that Greta Thunberg started), Greenpeace. And here are some social justice organizations: Uprose, and We Act.

There are also organizations that are writing climate policy in a new way and lobbying on Capitol Hill to get them passed. They are Citizen Climate Lobby, Climate Power, and Evergreen Action. Donate to them.

Or you can donate to groups that are working on electoral politics directly, like the Environmental Voter Project [in the United States]. The ability to put your preferred candidates in office is a huge part of the climate fight.

4. **Organize your workplace**—to ask your company to make greener business decisions or to lobby Congress for climate policies.

5. **And then finally, one of the most impactful things you can do is simply talk about climate change in your social networks**—especially when it feels most socially awkward and embarrassing.

Elissa Epel's To-Do List

This one is from Elissa Epel, the behavioral scientist, resilience expert, and my friend. Some of these tips are described in her book *The Stress Prescription*.[37] Elissa calls this list "Stress Resilience and Hope in the Anthropocene: Three Tips for Mental Preparedness," and says it is based on what she has learned from many others and inspired by her personal experience: "I have experienced deep climate distress, mental blocks on climate projects, and am sometimes visited by dark thoughts about how the miracle of our life on earth might end. While I feel the natural growing deep concern for our future, I have developed a sense of robust hope into action, and a way of living with the dialectic with more equanimity."

1. **Pack lightly!**
Our bags of luggage, both mental and physical, are too heavy and it's time to simplify, shed, and repack. Focus only on what matters to you most. Make that list of what you truly need in life and how you want to spend your limited time. A short list will allow you to live more lightly on earth, materially and spiritually, and to focus more time on living your purpose. Encourage others to follow you, by your joyous example.

2. **Welcome uncertainty.**
Expect the unexpected. We have both the usual uncertainty in our personal lives and now we have the volatile uncertainty of our unstable climate. The only thing we can be certain about in the future is uncertainty! With a sigh and a smile, soften your grip on the future. Indulge in the certainty of this present moment, breathing into this body with this air we share. People who appreciate what they already have feel the most life satisfaction and happiness. Find joy in what you do have now.

3. **Build hope, both absolute and active.**
Absolute hope: Hope is the foundation of stress resilience. Sometimes we hope for a specific outcome and our hopes are dashed. Feel into absolute hope, the robust hope that cannot die, regardless of what unfolds, that brings us together to keep doing our part to heal this world. What gives you hope? Can you feel the hope from the strength of our ancestors that brings us life today? The natural compassion humans are born with? The acts of altruism or generosity you witness? Hope and awe in the resilience of nature? Choose something that bolsters your sense of absolute hope for our future. Look for role models.

Active hope is hope infused with caring action, as Joanna Macy has described. Active hope cannot easily be lost. It is a highly contagious emotion because it models inspiring action. It is what spreads into social change. Remind yourself that active hope—doing the acts that are meaningful to you—brings the most powerful relief from pain, sorrow, anxiety, and anger.

You can create a vow of active hope that speaks to you. Here is one from Joanna Macy: "I vow to myself and each of you, to live on Earth more lightly and less violently in the food, products, and energy I consume."

And from me, Elissa: "I vow to myself and to each of you, to speak up when I see injustice to Earth beings and nature, and to find my own effective role in carbon reduction and my place in the network of social change."

Social influence transmits and sustains active hope, and the science of happiness reveals that joy is created most easily through social connection. Ask someone with common interests to create a resilience pod with you, a safe place you can share sadness, despair, hope, and support, and develop action.

Dr. Jane Goodall's To-Do List

The primatologist and anthropologist Jane Goodall has been one of my heroes since childhood, and her recent "MasterClass" includes several suggestions.[38]

▶ Make responsible decisions, realizing it's not just me, "it's hundreds of people, thousands of people, millions of people, eventually, hopefully, billions of people," including people like CEOs who can make big change with a single decision.

- Act locally. Think locally, too.
- Use the power of your dollar.
- Get the facts.
- Reconnect with nature.
- Communicate. Speak from the heart with a purpose. Have a sense of humor. Don't argue. Tell stories. Ask questions. And listen.
- Include girls and women. Listen to them. Educate them. Make sure schools have private bathroom facilities and hygiene products for girls when they're menstruating. Lend money to women. Let them plan their families.
- Parents and caretakers, support the interests and passions of children.
- Governments, invest in education, including early childhood education.
- Address the needs of people and nonhuman beings together. Work with local communities. It's not just about saving the chimps; it's also about improving the human communities around the forests where the chimps live. We need each other.
- Fund environmental projects and organizations.
- Teach the next generations to take care of the planet. For example, through youth programs like Roots & Shoots.

The Thunberg-Ernman Family's To-Do List

As Greta said in conversation with the Dalai Lama, "We now need to do everything we possibly can." In the book she wrote with her family, they offer the following actionable list (I added the bullets, but the points are theirs), which they introduce like this: "There already is a new story. And it is so positive that the angels are singing hallelujah and doing somersaults across the sky—

because we've already solved the climate crisis and we know the solutions are going to work. The solutions are so brilliant that they'll solve a great number of other problems all at once, such as growing wealth gaps, mental illness and inequality between the sexes. With the proviso that these solutions require fundamental changes and a concession or two in return."

- A very high CO_2 tax
- An overall goal of reducing greenhouse-gas emissions
- Planting enormous quantities of trees while allowing most of the existing forests to remain
- Slowing down and living on a smaller scale, collectively and locally, from local democracy to more collectively owned energy and food production
- Cooperating, because collective problems require collective solutions
- Investing in wind and solar energy instead of subsidizing fossil fuels
- Investing in existing technology instead of waiting for things that might come later, once it's already too late
- Changing many of our habits and many of us having to take a few ecological steps back
- Companies that have created the problems paying for everything they have messed up—companies that, even though they knew the risks, have seen unbelievable earnings as a result of destroying the climate and our ecosystems[39]

THE BEGINNING OF THE BEGINNING

Have you seen that movie *Don't Look Up*? It's a metaphor for our response to the climate crisis and—spoiler if you haven't seen it—it doesn't have a happy ending. As the movie's writer and director Adam McKay explains, this wasn't so much an ultimate prediction of doom as a realization that standard-fare Hollywood happy endings could be part of the problem. "I was just kicking around this idea—and part of it came from reading *Sapiens* by Yuval [Noah] Harari," Adam recollects. Summing up the thesis of that book, he says that it argues that what separates us, *Homo sapiens*, from our Neanderthal and Cro-Magnon ancestors is our ability to create myths and tell stories. "That got me thinking about what that means, that stories are that important. We've watched ten thousand movies—whether it's Marvel, James Bond, an action movie, *The Fast and the Furious*, the comedies, the stuff I've done—and it's always a happy ending. You just know it's coming. You know Hollywood's going to give it to you. I started wondering, are we sitting back and watching the climate and expecting a happy ending?"[40]

There's the tech-hero version of this happy-ending trap, when someone like Elon Musk says, "Technology will solve it." (Adam says "DiCaprio" told him that Elon Musk had said exactly that to "DiCaprio.") Kate Aronoff and others point to a literary white-male version of it, too, a phenomenon of certain famous male authors using their giant platforms to vault over the long, hard work of dedicated climate journalists to say, "Here's what *I* think we should do about the climate crisis."[41] These writers are not promising happiness, necessarily—in fact, they are quite sad—but in assuming solo authority they do give off heroic vibes. Kate and others like Rebecca Solnit, Mary Annaïse Heglar, and Amy Westervelt[42] say such climate heroes—brooding, delusional, or otherwise—are not what we need; what we do need is all of us taking back power from those who benefit most from the current status quo.

Greta, too, has been made a hero, though not willingly. In conversation with the Dalai Lama, when she felt the moderator was affording her too much praise, she was quick to correct her and share the credit with everyone in the movement. She would be the last person to say she alone can save us—or anyone, alone, can save us. That day with the Dalai Lama, Greta left us with a call to action: "We need to create a social movement. We need to shift the social norms, because if we are enough people who demand change, and who are advocating for these things, then we reach a critical mass and then we will no longer be possible to ignore, so that's what we need to do right now. It's not a small task, but it's something that we need to do because there is simply no other option." In the few years between Greta's first climate strike and the COVID-19 pandemic, she was doing just that, creating a social movement—or she would say, *we* were doing just that. And we were just getting going.

Adam McKay says in focus groups for *Don't Look Up*, the audience was very clear, "They said, 'We're sick of bullshit endings.'" I suspect the lack of bullshit—and the critique of bullshit—accounts, to a great extent, not only for the popularity of the movie but could apply also to the popularity of Greta herself. From her first posts on social media to her TED talk and speeches, we were struck by her honesty. When Greta speaks, she isn't playing games or pandering to her audiences, and she isn't in it for herself. In my favorite line in *Don't Look Up*, "DiCaprio's" character, an astrophysicist dragged into the political and media fray says, "Sometimes we need to just be able to say things to one another. We need to be able to hear things." By "things," he means the truth.

WHAT I WANT TO SAY IS that I realized in this conversation—with activists, scholars, and scientists; monks and moviemakers; teachers, writers, and journalists; with my friends and family; with the Dalai Lama, Greta, and you—that as scary, messy, and ugly as this crisis is, strangely, I wouldn't want to miss it. And I am glad, since there's no getting around it, to go through it with you, my fellow humans, whatever future we make for ourselves. I know someone who returned to New York City in 2001 after the first tech bubble burst—he had gone to San Francisco to work for a now-defunct tech news magazine and came back, as it happened, just in time for 9/11. He said in retrospect that as a New Yorker, he needed to be there when that happened, to go through it with his city. I know I'm not alone in feeling this way about this moment in human history, that it's an honor to be alive right now, to go through this with our Earth, to matter so much, and to be able to say things to one another and do things together that could heal and save us.

But in the spirit of listening, I'll give the last word to the young people struggling to love their future. This is not an ending. It is, as Greta says, "just the beginning of the beginning."[43]

EPILOGUE

THE DAY AFTER I DELIVERED THE MANUSCRIPT for this book to its publisher, Shambhala, US President Joseph Robinette Biden signed into law the Inflation Reduction Act, including the biggest piece of legislation ever enacted to address the climate crisis. The House had passed the bill, H.R. 5376, the Friday before and, on Monday, I pressed send, with tightly crossed fingers, on an email that had the manuscript attached.

Over the following weeks I followed the conversation, and I admit to worrying that a shoe was going to drop, that some catastrophic concession to the fossil-fuel industry would be revealed or some other way that the bill's center would not hold. This didn't happen. As far as I can tell, and according to the assessments of climate experts and journalists I trust, the investments, incentives, and greenhouse gas reductions in the Inflation Reduction Act will be transformative. Is the bill perfect or as big as it could have been? No. Is it huge and true to the original intentions of the many dedicated climate advisors who wrote it and activists who shaped it? Yes. Does it put us many significant steps down the road in the right direction? Yes. Was it a political miracle that any of it passed? Yes, and amen. To my mind, the most important caveat is not what's missing from this bill—we can get to that—but the fact that in a single election, it could all be undone. May we all—regardless of our political preferences, all of us who care about the future of humanity and our planet—not let that happen.

Around the same time, a friend of a friend was in Iceland and told me about a glacier there called Ok (in Icelandic, Okjökull), northeast of Reykjavik along the Kaldidalur valley road. *Kaldida-lur* means "cold valley," but the valley is less cold these days and the Ok glacier is not okay. In 2019, it disappeared completely. In its place now is the world's first memorial to a glacier lost to climate change, a plaque bearing these words by the author Andri Snær Magnason:

> Ok is the first Icelandic glacier to lose its status as a glacier. In the next 200 years all our glaciers are expected to follow the same path. This monument is to acknowledge that we know what is happening and what needs to be done. Only you know if we did it.

H.R. 5376 is a start—an ambitious start worth celebrating. Best case, it kicks off economic, political, and cultural feedback loops whereby things keep getting greener and the planet gets cooler. But only the future generations will know if we kept going in the right direction.

—September 1, 2022

ACKNOWLEDGMENTS

BOUNDLESS AND HEARTFELT GRATITUDE to the many people whose vision, guidance, and inspiration contributed directly and indirectly to this book. Indeed, there were many causes, conditions, and connections that brought it to fruition. While it's simply not possible to acknowledge everyone, there are some I'd like to lift up as they made a significant difference:

First and foremost, His Holiness the Dalai Lama and Greta Thunberg, generations and continents apart, for being bright beacons in this dark climate crisis. Their conversation inspired this book, and they continue to inspire me and millions more around the globe to wake up and show up every day.

Barry Hershey, a true bodhisattva, who had an aha moment about climate feedback loops that led to the creation of the Climate Emergency: Feedback Loops films, produced in collaboration with talented filmmakers Bonnie Waltch and Susan Gray. The films catalyzed the Conversation with His Holiness and Greta and, ultimately, this book.

Diana Chapman Walsh, chief architect and moderator of the Conversation, for being a role model and mentor to me and countless other women leaders.

Climate scientists Sue Natali and Bill Moomaw for being part of the Conversation, along with the many scientists in the films who contributed to the conversation in this book. Their lifelong dedication to climate research and advocacy is laudable.

Stephanie Tade, who from day one believed in this book and envisioned what it could be. And Stephanie Higgs, cowriter extraordinaire, who not only got the book but got me. It has been pure joy working with both Stephanies. We share chemistry and mutual care, unanticipated gifts from our work on this project.

Thupten Jinpa for being a valued thought partner. He suggested the structure for this book and played a critical role in interpreting nuance in His Holiness' teachings.

Daniel Donner and Sigrid Stavnem, helpful intermediaries, for their efforts to ensure clarity and integrity.

Nikko Odiseos, Matt Zepelin, Breanna Locke, and the whole team at Shambhala Publications for their vision and guidance every step of the way. And to Bernhard Salomon, Sophia Volpini de Maestre, and the team at *edition a* in Austria for initiating the first publication, *Kreislaufe des Klimawandels*, and recognizing the potential of this reimagined book.

Jacob Freund for his administrative support including securing the many permissions for this book. And to the entire staff at the Mind & Life Institute whose care and synergy are unparalleled.

The Mind & Life Summer Research Institute 2021 program planning committee—Elissa Epel, Bobbi Patterson, Dekila Chungyalpa, and Bruce Barrett—for their brilliance in creating an outstanding program with stellar scholars and activists, many of whom are interlocutors in this book.

The writing contributors—Willa Blythe Baker, Diana Beresford-Kroeger, Yuria Celidwen, Dekila Chungyalpa, Lyla June, Kritee Kanko, and Steve Leder—for their insightful words.

And finally, the many young activists who speak truth to power and keep insisting on a future they can love.

RESOURCES

APP OR WEB-BASED RESOURCES

350.org

All We Can Save Project—Resources for Educators
www.allwecansave.earth/for-educators

Better World Shopper
https://betterworldshopper.org

Climate Emergency: Feedback Loops—Educational Curriculum and Discussion Guides
https://feedbackloopsclimate.com/educational-materials

Connecting Data to Storytelling (Yale Program on Climate Change Communication)
https://climatecommunication.yale.edu/for-educators/connecting-data-to-storytelling

Earth.org

Earth4All
https://earth4all.life

Ethical Consumer
www.ethicalconsumer.org

Living Room Conversations—Climate Change Conversation Guide
https://livingroomconversations.org/topics/climate-change

Project Drawdown
https://drawdown.org

Reconnection: Meeting the Climate Crisis Inside Out
www.themindfulnessinitiative.org/reconnection

Regeneration
https://regeneration.org

"Ten Love Letters to the Earth" (by Thich Nhat Hanh)
https://emergencemagazine.org/essay/ten-love-letters-to-the-earth

The Mind, the Human-Earth Connection and the Climate Crisis
(online course)
www.mindandlife.org/climate-online-course

We Don't Have Time
https://app.wedonthavetime.org

BOOKS

Active Hope: How to Face the Mess We're in with Unexpected Resilience and Creative Power (Revised Edition) by Joanna Macy and Chris Johnstone

All We Can Save: Truth, Courage, and Solutions for the Climate Crisis, edited by Ayana Elizabeth Johnson and Katharine K. Wilkinson

Blessed Unrest: How the Largest Social Movement in History Is Restoring Grace, Justice, and Beauty to the World by Paul Hawken

Braiding Sweetgrass: Indigenous Wisdom, Scientific Knowledge, and the Teachings of Plants by Robin Wall Kimmerer

Climate: A New Story by Charles Eisenstein

Don't Even Think About It: Why Our Brains Are Wired to Ignore Climate Change by George Marshall

Earth for All: A Survival Guide for Humanity by Sandrine Dixson-Declève, Owen Gaffney, Jayati Ghosh, Jorgen Randers, Johan Rockström, and Per Espen Stoknes

Facing the Climate Emergency: How to Transform Yourself with Climate Truth (Second Edition) by Margaret Klein Salamon

Finding the Mother Tree: Discovering the Wisdom of the Forest by Suzanne Simard

The Future We Choose: Surviving the Climate Crisis by Christiana Figueres and Tom Rivett-Carnac

Generation Dread: Finding Purpose in an Age of Climate Anxiety by Britt Wray

Great Tide Rising: Towards Clarity and Moral Courage in a Time of Planetary Change by Kathleen Dean Moore

Limits to Growth: The 30-Year Update by Donella Meadows, Jorgen Randers, and Dennis Meadows

Oneness vs. the 1%: Shattering Illusions, Seeding Freedom by Vandana Shiva

Regeneration: Ending the Climate Crisis in One Generation by Paul Hawken

Saving Us: A Climate Scientist's Case for Hope and Healing in a Divided World by Katharine Hayhoe

Under a White Sky: The Nature of the Future by Elizabeth Kolbert

We Were Made for These Times: Ten Lessons for Moving Through Change, Loss, and Disruption by Kaira Jewel Lingo

You Matter More Than You Think: Quantum Social Change for a Thriving World by Karen O'Brien

Zen and the Art of Saving the Planet by Thich Nhat Hanh

FILMS AND VIDEOS

2040

An Inconvenient Truth

Breaking Boundaries

Climate Emergency: Feedback Loops (short films, feedback-loopsclimate.com)

David Attenborough: A Life on Our Planet

Don't Look Up

Earth Emergency

Eating Our Way to Extinction

"How Empowering Women and Girls Can Help Stop Global Warming" (Katharine Wilkinson TED talk)

Joanna Macy and the Great Turning

Taking Root: The Vision of Wangari Maathai

"The Dalai Lama with Greta Thunberg and Leading Scientists: A Conversation on the Crisis of Climate Feedback Loops"

"The Disarming Case to Act Right Now on Climate Change" (Greta Thunberg TEDx talk)

This Changes Everything

JOURNALISM

The Atlantic ("Planet")
www.theatlantic.com/projects/planet

Covering Climate Now
www.coveringclimatenow.org

Heated
https://heated.world/about

The Nation ("Climate Justice")
www.thenation.com/climate-change

The New Republic ("Climate")
www.newrepublic.com/climate

The New York Times ("Climate and Environment")
www.nytimes.com/section/climate

The New Yorker ("climate change" tag)
www.newyorker.com/tag/climate-change

ProPublica ("Environment")
www.propublica.org/topics/environment

The Uproot Project
www.grist.org/uproot

Volts
www.volts.wtf/about

ORGANIZATIONS AND INITIATIVES

The All We Can Save Project
www.allwecansave.earth

Bioneers
https://bioneers.org

The Carbon Underground
https://thecarbonunderground.org

Citizens' Climate Lobby
https://citizensclimatelobby.org

Council on the Uncertain Human Future
https://councilontheuncertainhumanfuture.org

Count Us In
https://count-us-in.com

Daughters for Earth
https://daughtersforearth.org

Fridays for Future
https://fridaysforfuture.org

GreenFaith
https://greenfaith.org

Humans & Nature
https://humansandnature.org

Inner Development Goals
www.innerdevelopmentgoals.org

The Mother Tree Project
https://mothertreeproject.org

Ocean Optimism
www.oceanoptimism.org

One Earth Sangha
https://oneearthsangha.org

One Resilient Earth
www.oneresilientearth.org

Our Climate
https://ourclimate.us

Planetary Guardians
https://planetaryguardians.global

Planetary Health Alliance
https://planetaryhealthalliance.org

Sunrise Movement
www.sunrisemovement.org

Work That Reconnects
https://workthatreconnects.org

PODCASTS

Climate One

The Climate Pod

Emergence Magazine

Global Weirding with Katharine Hayhoe (NPR series)

How to Save a Planet

"The Human-Earth Connection: Interview with Dekila
    Chungyalpa," *Mind & Life*

In the Deep with Catherine Ingram

A Matter of Degrees

"People, Policy, and Planet: An Interview with Jamie Bristow,"
    Mind & Life

Scene on Radio (season 5)

TED Climate

Volts

"A Wild Love for the World: An Interview with Joanna Macy,"
    On Being

INTERLOCUTORS

Kate Aronoff
www.newrepublic.com/authors/kate-aronoff
Twitter: @katearonoff

Willa Blythe Baker
www.shambhala.com/authors/a-f/willa-blythe-baker.html

Camille Barton
www.camillebarton.co.uk

Stephen Batchelor
www.stephenbatchelor.org
Instagram: @agnostic108

Diana Beresford-Kroeger
www.dianaberesford-kroeger.com

Dekila Chungyalpa
www.centerhealthyminds.org/programs/loka-initiative
Twitter: @dchungyalpa

Mike Coe
www.woodwellclimate.org/staff/michael-coe
Twitter: @WoodwellClimate

The Dalai Lama
www.dalailama.com
Twitter: @DalaiLama

Phil Duffy
www.woodwellclimate.org/staff/philip-duffy
Twitter: @WoodwellClimate

Kerry Emanuel
https://emanuel.mit.edu/

Elissa Epel
https://profiles.ucsf.edu/elissa.epel
Twitter: @Dr_Epel

Malena Ernman
Instagram: @malena_ernman

Christiana Figueres
www.christianafigueres.com
Twitter: @CFigueres

Andy Fisher
www.andyfisher.ca
Twitter: @FisherTrueLine

Jennifer Francis
www.jenniferafrancis.com
Twitter: @JFrancisClimate

Charlotte Gill
www.charlottegill.com
Twitter: @charlotte_gill

Jane Goodall
www.janegoodall.org
Twitter and Instagram: @janegoodallinst

Genevieve Guenther
www.genevieveguenther.com
Twitter: @DoctorVive

Joan Halifax
www.upaya.org
Twitter: @jhalifax

Carl Hall
www.westoneint.com/team

Paul Hawken
www.paulhawken.com
Twitter: @PaulHawken

Katharine Hayhoe
www.katharinehayhoe.com
Twitter: @KHayhoe

Barry Hershey
www.barryhershey.com

Marika Holland
https://staff.ucar.edu/users/mholland

Catherine Ingram
www.catherineingram.com
Twitter: @CathIngram

Thupten Jinpa
www.mindandlife.org/person/thupten-jinpa

Ayana Elizabeth Johnson
www.ayanaelizabeth.com
Twitter: @ayanaeliza

Lyla June
www.lylajune.com
Twitter: @lylajune

Kritee Kanko
www.boundlessinmotion.org
Twitter: @KriteeKanko

Riane Konc
www.rianekonc.com
Twitter: @theillustrious

Francesco Lastrucci
www.francescolastrucci.com
Instagram: @francescolastrucci

Beverly Law
https://directory.forestry.oregonstate.edu/people/law-beverly

Steve Leder
www.steveleder.com
Twitter: @Steve_Leder

David Loy
www.davidloy.org

Wangari Maathai
www.greenbeltmovement.org/wangari-maathai

Joanna Macy
www.joannamacy.net

Ed Maibach
www.climatechangecommunication.org
Twitter: @MaibachEd

Andrew Marantz
Twitter: @andrewmarantz

Adam McKay
www.imdb.com/name/nm0570912
Twitter: @GhostPanther

Donella Meadows
www.donellameadows.org

George Monbiot
www.monbiot.com
Twitter: @GeorgeMonbiot

William (Bill) Moomaw
www.earthwatch.org/scientists/william-moomaw-phd

Anupam Nanda
www.anupam-nanda.org

Susan Natali
www.woodwellclimate.org/staff/susan-natali
Twitter: @woodwellarctic

Karen O'Brien
www.youmattermorethanyouthink.com/about

Jenny Odell
www.jennyodell.com
Twitter: @the_jennitaur

Barbara (Bobbi) Patterson
www.mindandlife.org/person/bobbi-patterson

Donald (Don) Perovich
www.icedrill-education.org/personnel/don-perovich

Matthieu Ricard
www.matthieuricard.org
Twitter: @MatthieuRicard

Andrew Rice
www.andrewrice.net
Twitter: @riceid

Brendan Rogers
www.woodwellclimate.org/staff/brendan-rogers
Twitter: @woodwellarctic

Nick Romeo
www.nickromeowriter.com
Twitter: @Nickromeoauthor

Regina Romero
Twitter: @TucsonRomero

Jonathan Rose
www.garrisoninstitute.org/person/jonathan-f-p-rose
Twitter: @JonathanFPRose

Sebene Selassie
www.sebeneselassie.com

Vandana Shiva
www.navdanyainternational.org
Twitter: @drvandanashiva

Rebecca Solnit
www.rebeccasolnit.net
Twitter: @RebeccaSolnit

Greta Thunberg
Twitter: @GretaThunberg
Instagram: @gretathunberg

Bonnie Waltch
Twitter: @BonnieWaltch

Christine Wamsler
www.lucsus.lu.se/christine-wamsler

Warren Washington
www.cgd.ucar.edu/staff/wmw

Kyle Whyte
https://seas.umich.edu/research/faculty/kyle-whyte
Twitter: @kylepowyswhyte

Katharine K. Wilkinson
www.kkwilkinson.com
www.allwecansave.earth
Twitter: @DrKWilkinson

George Woodwell
www.woodwellclimate.org/staff/george-woodwell
Twitter: @WoodwellClimate

Adriana Zuniga
https://las.arizona.edu/people/adriana-zuniga-teran

NOTES

INTRODUCTION

1. 1.5 or 2 degrees Celsius equals 2.7 or 3.6 degrees Fahrenheit.
2. Dalai Lama, *His Holiness the 14th Dalai Lama on Environment: Collected Statements 1987–2017*, 6th ed. (Dharamsala: Environment and Policy Desk, Tibet Policy Institute, Central Tibetan Administration, 2017), 88.
3. Dalai Lama, *On Environment*, 89.
4. Dalai Lama, 89–90.
5. Dalai Lama, 165.
6. Baher Kamal, "Climate Migrants Might Reach One Billion by 2050," Reliefweb, August 21, 2017, https://reliefweb.int/report /world/climate-migrants-might-reach-one-billion-2050.
7. Dalai Lama, *On Environment*, 164–65.
8. Greta Thunberg, *No One Is Too Small to Make a Difference* (New York: Penguin Books, 2021), 66, Kindle.
9. Greta Thunberg, Svante Thunberg, Malena Ernman, Beata Ernman, *Our House Is on Fire: Scenes from a Family and a Planet in Crisis*, trans. Paul Norlen and Saskia Vogel (New York: Penguin Books, 2020), 207–8.
10. Thunberg, *No One Is Too Small*, 10.
11. Council on the Uncertain Human Future, https:// councilontheuncertainhumanfuture.org.
12. David Marchese, "Yuval Noah Harari Believes This Simple Story Can Save the Planet," *New York Times Magazine*, November 7, 2021, www.nytimes.com/interactive/2021/11/08 /magazine/yuval-noah-harari-interview.html.
13. Thunberg, *No One Is Too Small*, 2.

I. THE SCIENCE: WHY THE ICE, WIND, CLOUDS, AND TREES MATTER

1. Best I can make out from hearing her pronounce it. Greta generously says she doesn't mind all the different ways people say her name: "There's no wrong way to pronounce it. Everyone pronounces it in their own way." In an interview with Amy Goodman, *Democracy Now!*, September 11, 2019, YouTube, https://youtu.be/Dgi3oWy_V74.
2. Richard A. Houghton and George M. Woodwell, "Global Climatic Change," *Scientific American* 260, no. 4 (April 1989): 36.
3. Climate Emergency: Feedback Loops, directed by Susan Gray (Boston: Northern Light Productions, 2021), https://feedbackloopsclimate.com/. All the scientists in this chapter are quoted from the *Climate Emergency* films.
4. Greta Thunberg, *No One Is Too Small to Make a Difference* (New York: Penguin Books, 2021), 96–97, Kindle.
5. Thunberg, *No One Is Too Small*, 19, 60, 76, 79, 97, 106.
6. Andrew Rice, "This Is New York in the Not-So-Distant Future," *New York Magazine*, September 5, 2016, https://nymag.com/intelligencer/2016/09/new-york-future-flooding-climate-change.html.
7. Sarah Miller, "The Millions of Tons of Carbon Emissions That Don't Officially Exist," *The New Yorker*, December 8, 2021, www.newyorker.com/news/annals-of-a-warming-planet/the-millions-of-tons-of-carbon-emissions-that-dont-officially-exist.
8. Charlotte Gill, *Eating Dirt: Deep Forests, Big Timber, and Life with the Tree-Planting Tribe* (Vancouver: Greystone Books, 2011), 158–59.
9. Donella Meadows, Jorgen Randers, and Dennis Meadows, *Limits to Growth: The 30-Year Update* (White River Junction, VT: Chelsea Green, 2004), 493.
10. Meadows, Randers, Meadows, *Limits to Growth*, 494.

2. THE SPIRIT: THE PROBLEM WITH BUSINESS AS USUAL

1. Quoted in David Loy, "Indra's Postmodern Net," *Philosophy East & West* 43, no. 3 (July 1993): 481.
2. Dalai Lama, *His Holiness the 14th Dalai Lama on Environment: Collected Statements 1987–2017*, 6th ed. (Dharamsala: Environment and Policy Desk, Tibet Policy Institute, Central Tibetan Administration, 2017), 54–55.
3. Simon Hattenstone, "The Transformation of Greta Thunberg," *The Guardian*, September 25, 2021, www.theguardian.com /environment/ng-interactive/2021/sep/25/greta-thunberg -i-really-see-the-value-of-friendship-apart-from-the-climate -almost-nothing-else-matters.
4. Greta Thunberg, *No One Is Too Small to Make a Difference* (New York: Penguin Books, 2021), 106.
5. Thunberg, *No One Is Too Small*, 96.
6. Anthony Leiserowitz et al., "Climate Change in the American Mind: November 2019," Yale Program on Climate Change Communication, December 17, 2019, https:// climatecommunication.yale.edu/publications/climate-change -in-the-american-mind-november-2019/toc/2/.
7. Hattenstone, "Transformation of Greta Thunberg."
8. Raymond Zhong, "These Climate Scientists Are Fed Up and Ready to Go on Strike," *New York Times*, March 1, 2022, www .nytimes.com/2022/03/01/climate/ipcc-climate-scientists -strike.html.
9. Donella Meadows, "Leverage Points: Places to Intervene in a System," The Donella Meadows Project, accessed August 15, 2022, https://donellameadows.org/archives/leverage-points -places-to-intervene-in-a-system/.
10. "Greenhouse 100 Polluters Index (2021 Report, Based on 2019 Data)," Political Economy Research Institute, accessed August 15, 2022, https://peri.umass.edu/greenhouse-100-polluters -index-current.

11. Dalai Lama, *On Environment*, 26.
12. David Loy, "Is the Ecological Crisis Also a Spiritual Crisis?" Buddhism and Ecology Summit, Tricycle Online Courses, April 18, 2022, https://learn.tricycle.org/p/the-buddhism-and -ecology-summit.
13. *Joanna Macy and the Great Turning*, directed by Chris Landry (2014), short film, 27 min.
14. Thunberg, *No One Is Too Small*, 96.
15. Donella Meadows, Jorgen Randers, and Dennis Meadows, *Limits to Growth: The 30-Year Update* (White River Junction, VT: Chelsea Green, 2004), 490–91.
16. Thunberg, *No One Is Too Small*, 2.
17. Stephen Batchelor, "Embracing Extinction," Buddhism and Ecology Summit, Tricycle Online Courses, April 18, 2022, https://learn.tricycle.org/p/the-buddhism-and-ecology -summit.
18. Donella Meadows, "Envisioning a Sustainable World" (lecture, Third Biennial Meeting of the International Society for Ecological Economics, San Jose, Costa Rica, October 24–28, 1994); The Donella Meadows Project, https:// donellameadows.org/archives/envisioning-a-sustainable -world/.
19. Robert B. Jackson et al., "Human Well-Being and Per Capita Energy Use," *Ecosphere* 13, no. 4 (April 2022), https:// esajournals.onlinelibrary.wiley.com/doi/10.1002/ecs2.3978.
20. Upaya Zen Center, "Ending the Smog of Ignorance," Upaya's blog, July 3, 2022, https://www.upaya.org/2022/07/ridding -the-smog-of-ignorance/.
21. *Joanna Macy and the Great Turning*.
22. Compiled from Joanna Macy and Molly Brown, *Coming Back to Life: The Updated Guide to the Work That Reconnects* (Gabriola Island, BC: New Society, 2014), 30–31, and from *Joanna Macy and the Great Turning*.
23. Loy, "Is the Ecological Crisis Also a Spiritual Crisis?"

3. EARTH'S CAPACITY: LET THE EARTH DO WHAT THE EARTH DOES

1. Paul Hawken, "Meaningful Action to Achieve Planetary Health," Buddhism and Ecology Summit, Tricycle Online Courses, April 21, 2022, https://learn.tricycle.org/p/the-buddhism-and-ecology-summit.
2. Climate Emergency: Feedback Loops, directed by Susan Gray (Boston: Northern Light Productions, 2021), https://feedbackloopsclimate.com/.
3. Dalai Lama, *His Holiness the 14th Dalai Lama on Environment: Collected Statements 1987–2017*, 6th ed. (Dharamsala: Environment and Policy Desk, Tibet Policy Institute, Central Tibetan Administration, 2017), 23.
4. Dalai Lama, *On Environment*, 29.
5. *Taking Root: The Vision of Wangari Maathai*, directed by Lisa Merton and Alan Dater (Marlboro, VT: Marlboro Productions, 2008), Vimeo, https://vimeopro.com/marlboroproductions/taking-root-the-vision-of-wangari-maathai.
6. Francesco Lastrucci, "Inside the Campaign to Save an Imperiled Cambodian Rainforest," *New York Times*, December 20, 2021, www.nytimes.com/2021/12/20/travel/cardamom-mountains-wildlife-cambodia.html.
7. Paul Hawken, *Blessed Unrest: How the Largest Movement in the World Came into Being and Why No One Saw It Coming* (New York: Viking, 2007), 12–18.
8. Hawken, "Meaningful Action."
9. Quinn McVeigh, "Tucson Is Planting a Million Trees to Combat Climate Change," Good Good Good, December 27, 2021, www.goodgoodgood.co/articles/tucson-million-trees-initiative.
10. Diana Beresford-Kroeger, *To Speak for the Trees: My Life's Journey from Ancient Celtic Wisdom to a Healing Vision of the Forest* (Toronto: Random House Canada, 2019), 183–87.
11. Quotes in this last paragraph are from Tracy L. Barnett and Lyla June Johnston, "Kelp Gardens, Piñion Forests: Lyla June on Renovating Foodways as a Path to Sovereignty," Esperanza

Project, November 9, 2019, www.esperanzaproject.com/2019
/environment/kelp-gardens-and-pinon-forests-native
-regenerative-agriculture/.

12. Kyle Powys White, "Conveners of Responsibilities," Center
for Humans and Nature, September 30, 2013, https://
humansandnature.org/earth-ethic-kyle-powys-whyte/.

13. Michael Mack, "Cities Are the Answer: Jonathan Rose Sees
Cities as Key to Sustainability," MIT Center for Real Estate,
accessed August 15, 2022, https://mitcre.mit.edu/news
/partnernews/cities-are-the-answer.

14. Nick Romeo, "How Oslo Learned to Fight Climate Change,"
The New Yorker, May 4, 2022, www.newyorker.com/news
/annals-of-a-warming-planet/how-oslo-learned-to-fight
-climate-change.

15. Anupam Nanda, "Superblocks; Barcelona's Car-Free
Zones Could Extend Lives and Boost Mental Health," The
Conversation, September 13, 2019, https://theconversation
.com/superblocks-barcelonas-car-free-zones-could-extend
-lives-and-boost-mental-health-123295.

16. Greta Thunberg, *No One Is Too Small to Make a Difference*
(New York: Penguin Books, 2021), 39, Kindle.

17. Jonathon Porritt, "Environmentalist Jonathon Porritt's Big Idea
to Slow Global Warming," *New Scientist*, August 18, 2021, www
.newscientist.com/article/2287520-environmentalist-jonathon
-porritts-big-idea-to-slow-global-warming.

18. Zeke Hausfather, "Let's Not Pretend Planting Trees Is a
Permanent Climate Solution," *New York Times*, June 4, 2022,
www.nytimes.com/2022/06/04/opinion/environment
/climate-change-trees-carbon-removal.html.

19. Dalai Lama, *On Environment*, 91–92. Emphasis mine.

20. His Holiness the Dalai Lama, *The Universe in a Single Atom:
The Convergence of Science and Spirituality* (New York: Morgan
Road Books, 2005), Conclusion: Science, Spirituality, and
Humanity, Kindle.

4. HUMAN CAPACITY: THE NECESSITY OF A SENSE OF EFFICACY

1. Greta Thunberg, "The Disarming Case to Act Right Now on Climate Change," TEDxStockholm, January 27, 2019, www.ted .com/talks/greta_thunberg_the_disarming_case_to_act _right_now_on_climate_change?language=en.
2. *Joanna Macy and the Great Turning*, directed by Chris Landry (2014), short film, 27 min.
3. Greta Thunberg, *No One Is Too Small to Make a Difference* (New York: Penguin Books, 2021), 102–4, Kindle.
4. Thunberg, *No One Is Too Small*, 33, 72, 106.
5. Dalai Lama, *His Holiness the 14th Dalai Lama on Environment: Collected Statements 1987–2017*, 6th ed. (Dharamsala: Environment and Policy Desk, Tibet Policy Institute, Central Tibetan Administration, 2017),114–15.
6. Thunberg, *No One Is Too Small*, 5.
7. Thunberg, 33.
8. Thupten Jinpa, *A Fearless Heart: How the Courage to Be Compassionate Can Transform Our Lives* (New York: Avery, 2015).
9. Dalai Lama, *On Environment*, 53.
10. Thunberg, *No One Is Too Small*, 12.
11. Thunberg, 4.
12. Thunberg, 72.
13. Rebecca Solnit, *Men Explain Things to Me* (Chicago: Haymarket Books, 2014), chapter 6, Kindle.
14. Dalai Lama, *On Environment*, 81.
15. Thunberg, *No One Is Too Small*, 40.

5. HEARTBREAK: THE DARKNESS AND THE LIGHT

1. Joanna Macy quoted in Susanne Moser, "To Behold Worlds Ending," in *A Wild Love for the World: Joanna Macy and the Work of Our Time,* ed. Stephanie Kaza (Boulder, CO: Shambhala Publications, 2020), 84.

2. George Monbiot, "Watching *Don't Look Up* Made Me See My Whole Life of Campaigning Flash Before Me," *The Guardian,* January 4, 2022, www.theguardian.com/commentisfree/2022 /jan/04/dont-look-up-life-of-campaigning.

3. *Joanna Macy and the Great Turning,* directed by Chris Landry (2014), short film, 27 min.

4. Macy, *Wild Love,* 73.

5. Macy, 73–74.

6. Catherine Ingram, "Facing Extinction," catherineingram, 2019 (updated frequently), www.catherineingram.com /facingextinction/.

7. Monbiot, "Watching *Don't Look Up.*"

8. *Joanna Macy and the Great Turning.*

9. Andy Fisher, "The Thrumming Relationality of All Things," in *A Wild Love for the World: Joanna Macy and the Work of Our Time,* ed. Stephanie Kaza (Boulder, CO: Shambhala Publications, 2020), 91.

10. David Wallace-Wells, "Why Is the World Ignoring the Latest U.N. Climate Report?" *New York Magazine,* March 14, 2022, https://nymag.com/intelligencer/2022/03/un-climate-report .html.

11. "Global CO_2 Emissions Rebounded to Their Highest Level in History in 2021," press release, IEA (International Energy Agency), March 8, 2022, www.iea.org/news/global-co2 -emissions-rebounded-to-their-highest-level-in-history-in-2021.

12. Donald Trump and Greta Thunberg quoted in Simon Hattenstone, "The Transformation of Greta Thunberg," *The Guardian,* September 25, 2021, www.theguardian.com /environment/ng-interactive/2021/sep/25/greta-thunberg -i-really-see-the-value-of-friendship-apart-from-the-climate -almost-nothing-else-matters.

13. *I Am Not Your Negro*, directed by Raoul Peck (New York: Magnolia Pictures, 2016).

14. Macy, *Wild Love*, 75.

15. Macy, 76.

16. Daniel Menaker, *Terminalia: Poems* (Washington, DC: Portal Press, 2020), 24.

17. Camille Sapara Barton, "The GEN Grief Toolkit: Embodiment Tools and Rituals to Support Grief Work in Community," Global Environments Network, January 2022, https:// globalenvironments.org/toolkits/grief-toolkit/#1.

18. From a conversation on May 27, 2022.

19. Willa Blythe Baker, "The Alchemy of Despair," Buddhism and Ecology Summit, Tricycle Online Courses, April 21, 2022, https://learn.tricycle.org/p/the-buddhism-and-ecology -summit.

20. Willa Blythe Baker, *The Wakeful Body: Somatic Mindfulness as a Path to Freedom* (Boulder, CO: Shambhala Publications, 2021), 96–99.

21. Riane Konc, "Excerpts from the All-Girl Remake of 'Lord of the Flies,'" *The New Yorker*, September 7, 2017, www.newyorker .com/humor/daily-shouts/excerpts-from-the-all-girl-remake -of-lord-of-the-flies.

22. Ingram, "Facing Extinction."

23. Stephen Batchelor, *Buddhism without Beliefs: A Contemporary Guide to Awakening* (New York: Riverhead, 1997), 32.

24. Stephen Batchelor, "Embracing Extinction," Buddhism and Ecology Summit, Tricycle Online Courses, April 18, 2022, https://learn.tricycle.org/p/the-buddhism-and-ecology -summit.

25. Stephen Batchelor, "Embracing Extinction," *Tricycle: The Buddhist Review*, Fall 2020, https://tricycle.org/magazine /stephen-batchelor-climate.

26. Batchelor, "Embracing Extinction," *Tricycle*.

27. *Joanna Macy and the Great Turning*.

28. Steve Leder, *The Beauty of What Remains: How Our Greatest Fear Becomes Our Greatest Gift* (New York: Avery, 2021).

29. Steve Leder with Katie Couric, "Rabbi Steve Leder Discusses

the Mental Health Impact of World Events and Recent Tragedies," *Next Question with Katie Couric,* June 4, 2022, https://youtu.be/QH2LnnH-6PQ.

6. WONDERMENT: A PRESENT WE CAN LOVE, A FUTURE WE CAN IMAGINE

1. Jenny Odell, *How to Do Nothing: Resisting the Attention Economy* (Brooklyn: Melville House, 2019), 19, Kindle.
2. Jenny Odell, "Jenny Odell on How to Do Nothing," *Offline with Jon Favreau,* January 23, 2022, https://offline-with-jon-favreau .simplecast.com/episodes/jenny-odell-on-how-to-do-nothing.
3. Odell, *How to Do Nothing,* 4.
4. Odell, 19.
5. Odell, 22.
6. Gretchen Reynolds, "An 'Awe Walk' Might Do Wonders for Your Well-Being," *New York Times,* September 30, 2020, www .nytimes.com/2020/09/30/well/move/an-awe-walk-might -do-wonders-for-your-well-being.html.
7. When a meeting scheduled to be in Santiago was canceled by the Chilean government after climate demonstrations in the streets, the UN moved it on short notice to Madrid and Greta had to find a sailboat to take her back to Europe sooner than expected.
8. Tara Law, "Climate Activist Greta Thunberg, 16, Arrives in New York After Sailing Across the Atlantic," *Time,* August 28, 2019, https://time.com/5663534/greta-thunberg-arrives-sail-atlantic.
9. Odell, *How to Do Nothing,* 130–31.
10. Odell, 20.
11. Odell, "Jenny Odell on How to Do Nothing."
12. Sebene Selassie, "Just Ride," Sebene Selassie, accessed August 15, 2022, www.sebeneselassie.com/blog/just-ride.
13. Rebecca Solnit with Mary Annaïse Heglar and Amy Westervelt, "Down Uterus, Down Girl!" *Hot Take,* May 20, 2022, https:// hot-take.simplecast.com/episodes/down-uterus-down-girl.

14. Greta Thunberg, Svante Thunberg, Malena Ernman, and Beata Ernman, *Our House Is on Fire: Scenes from a Family and a Planet in Crisis*, trans. Paul Norlen and Saskia Vogel (New York: Penguin Books, 2020), 177.

15. Susan Bauer-Wu, *Leaves Falling Gently: Living Fully with Serious and Life-Limiting Illness through Mindfulness, Compassion and Connectedness* (Oakland, CA: New Harbinger, 2011), 3.

16. Wendell Berry, "The Peace of Wild Things," *The Selected Poems of Wendell Berry* (Berkeley, CA: Counterpoint, 2009), 41, Kindle.

17. Donella Meadows, "Envisioning a Sustainable World" (lecture, Third Biennial Meeting of the International Society for Ecological Economics, San Jose, Costa Rica, October 24–28, 1994); The Donella Meadows Project, https://donellameadows.org/archives/envisioning-a-sustainable-world/.

18. Meadows, "Envisioning a Sustainable World."

19. Meadows, "Envisioning."

20. Solnit, "Down Uterus," *Hot Take*.

21. Donella Meadows, Jorgen Randers, and Dennis Meadows, *Limits to Growth: The 30-Year Update* (White River Junction, VT: Chelsea Green, 2004), 482.

22. Meadows, Randers, Meadows, *Limits to Growth*, 482–83.

7. THE BEGINNING OF THE AGE OF ENOUGH: WHAT TO DO AND HOW TO THINK ABOUT DOING IT

1. Willa Blythe Baker, "The Alchemy of Despair," Buddhism and Ecology Summit, Tricycle Online Courses, April 21, 2022, https://learn.tricycle.org/p/the-buddhism-and-ecology-summit.

2. Greta Thunberg, Svante Thunberg, Malena Ernman, and Beata Ernman, *Our House Is on Fire: Scenes from a Family and a Planet in Crisis*, trans. Paul Norlen and Saskia Vogel (New York: Penguin Books, 2020), 242.

3. Greta Thunberg, *No One Is Too Small to Make a Difference* (New York: Penguin Books, 2021), 106, Kindle.

4. A tenet of the evidence-based approach to change called motivational interviewing, for example.

5. Andrew Marantz, "The Youth Movement Trying to Revolutionize Climate Politics," *The New Yorker*, February 28, 2022, www.newyorker.com/magazine/2022/03/07/the-youth -movement-trying-to-revolutionize-climate-politics.

6. Karen O'Brien, *You Matter More Than You Think: Quantum Social Change for a Thriving World* (Oslo: cCHANGE Press, 2021), 117.

7. O'Brien, *You Matter More Than You Think*, 120–21.

8. O'Brien, 38.

9. *Taking Root: The Vision of Wangari Maathai*, directed by Lisa Merton and Alan Dater (Marlboro, VT: Marlboro Productions, 2008), Vimeo, https://vimeopro.com/marlboroproductions /taking-root-the-vision-of-wangari-maathai.

10. Thunberg et al., *Our House*, 179.

11. Thunberg et al., 114.

12. Rebecca Solnit with Mary Annaïse Heglar and Amy Westervelt, "Down Uterus, Down Girl!" *Hot Take*, May 20, 2022, https://hot-take.simplecast.com/episodes/down-uterus -down-girl.

13. Thunberg et al., *Our House*, 151–52.

14. Thunberg et al., 176.

15. Dekila Chungyalpa, "At the Center of All Things Is Interdependence," Humans and Nature, May 17, 2021, https://humansandnature.org/at-the-center-of-all-things-is -interdependence/.

16. Dalai Lama, *His Holiness the 14th Dalai Lama on Environment: Collected Statements 1987–2017*, 6th ed. (Dharamsala: Environment and Policy Desk, Tibet Policy Institute, Central Tibetan Administration, 2017), 57.

17. Tracy L. Barnett and Lyla June Johnson, "Kelp Gardens, Piñon Forests," The Esperanza Project, November 9, 2019, www .esperanzaproject.com/2019/environment/kelp-gardens-and -pinon-forests-native-regenerative-agriculture/.

18. Quoted in Catherine Ingram, "Facing Extinction," catherineingram, 2019 (updated frequently), www .catherineingram.com/facingextinction/.

19. Thunberg et al., *Our House*, 157.

20. Sebene Selassie, "Connecting: Earth | Water | Fire | Air," Sebene Selassie, accessed August 15, 2022, www.sebeneselassie .com/blog/connecting-earth-water-fire-air.

21. Dalai Lama, *On Environment*, 35.

22. Jenny Odell, *How to Do Nothing: Resisting the Attention Economy* (Brooklyn: Melville House, 2019), 154, Kindle.

23. Kate Aronoff, "Things Are Bleak!," *The Nation*, October 29, 2019, www.thenation.com/article/archive/jonathan-safran -foer-we-are-the-weather-climate-review/.

24. Loran Nordgren with Shankar Vedantam, "Work 2.0," *Hidden Brain*, November 1, 2021, https://hiddenbrain.org/podcast /work-2-0-the-obstacles-you-dont-see/.

25. Jenny Odell, "Jenny Odell on How to Do Nothing," *Offline with Jon Favreau*, January 23, 2022, https://offline-with-jon-favreau .simplecast.com/episodes/jenny-odell-on-how-to-do-nothing.

26. Paul Hawken, *Regeneration: Ending the Climate Crisis in One Generation* (New York: Penguin Books, 2021), 250.

27. Seth Wynes and Kimberly A. Nicholas, "The Climate Mitigation Gap: Education and Government Recommendations Miss the Most Effective Individual Actions," *Environmental Research Letters* 12, no. 7 (July 12, 2017), https://iopscience.iop.org/article/10.1088/1748-9326/aa7541. As the study summarized their findings, "We recommend four widely applicable high-impact (i.e., low emissions) actions with the potential to contribute to systemic change and substantially reduce annual personal emissions: having one fewer child (an average for developed countries of 58.6 tonnes CO_2-equivalent (tCO_2e) emission reductions per year), living car-free (2.4 tCO_2e saved per year), avoiding airplane travel (1.6 tCO_2e saved per roundtrip transatlantic flight), and eating a plant-based diet (0.8 tCO_2e saved per year). These actions have much greater potential to reduce emissions than commonly promoted strategies like comprehensive recycling (four times

less effective than a plant-based diet) or changing household lightbulbs (eight times less). Though adolescents poised to establish lifelong patterns are an important target group for promoting high-impact actions, we find that ten high school science textbooks from Canada largely fail to mention these actions (they account for 4% of their recommended actions), instead focusing on incremental changes with much smaller potential emissions reductions. Government resources on climate change from the EU, USA, Canada, and Australia also focus recommendations on lower-impact actions. We conclude that there are opportunities to improve existing educational and communication structures to promote the most effective emission-reduction strategies and close this mitigation gap."

28. For example, Ezra Klein, "Your Kids Are Not Doomed," *New York Times*, June 5, 2022, www.nytimes.com/2022/06/05/opinion/climate-change-should-you-have-kids.html.

29. Genevieve Guenther and David Wallace-Wells with Jane Coaston, "Got Climate Doom? Here's What You Can Do to Actually Make a Difference," The Argument, *New York Times*, November 10, 2021, www.nytimes.com/2021/11/10/opinion/climate-change-personal-actions.html?showTranscript=1.

30. Jane Goodall, "Dr. Jane Goodall Teaches Conservation," MasterClass, accessed August 15, 2022, www.masterclass.com/classes/jane-goodall-teaches-conservation.

31. Hawken, *Regeneration*, 72.

32. Hawken, 249–50.

33. Ayana Elizabeth Johnson and Katharine K. Wilkinson, eds., *All We Can Save: Truth, Courage, and Solutions for the Climate Crisis* (New York: One World, 2020), xxii–xxiv.

34. Catherine Ingram, "Facing Extinction," catherineingram, 2019 (updated frequently), www.catherineingram.com/facingextinction/.

35. Donella Meadows, Jorgen Randers, and Dennis Meadows, *Limits to Growth: The 30-Year Update* (White River Junction, VT: Chelsea Green, 2004), 492–93.

36. Guenther and Wallace-Wells, "Got Climate Doom?"

37. Elissa Epel, *The Stress Prescription: 7 Days to More Joy and Ease* (New York: Penguin Life, 2022).

38. Goodall, "Dr. Jane Goodall Teaches Conservation."

39. Thunberg et al., *Our House*, 175–76.

40. Adam McKay with David Roberts, *Volts*, January 12, 2022, www.volts.wtf/p/volts-podcast-dont-look-up-director#details.

41. Aronoff, "Things Are Bleak!"

42. Solnit, "Down Uterus" *Hot Take*.

43. Thunberg, *No One Is Too Small*, 106.

CREDITS

Quoted material in this book from the following sources is included with kind permission:

Opening blessing copyright © by Yuria Celidwen.

"When Pessimism Is Not an Option" and "One Humanity Practice" copyright © by His Holiness the Dalai Lama.

"A Meditation for Ecoanxiety and Climate Despair" copyright © by Dekila Chungyalpa.

"Practicing with Heartbreak" copyright © by Kritee Kanko.

"What It Means to Sacrifice" copyright © by Steve Leder.

"Go To the Earth in the Morning" copyright © by Lyla June.

Excerpts from *No One Is Too Small to Make a Difference* by Greta Thunberg, copyright © 2018, 2019 by Greta Thunberg. Used by permission of Penguin Books, an imprint of Penguin Publishing Group, a division of Penguin Random House LLC. All rights reserved.

Excerpts from *Our House Is On Fire: Scenes of a Family and a Planet in Crisis* by Greta Thunberg, Svante Thunberg, Malena Ernman, and Beata Ernman, copyright © 2018 by Malena Ernman, Svante Thunberg, Greta Thunberg, Beata Ernman, and Bokförlaget Polaris translation copyright © 2020 by Paul Norlen and Saskia Vogel. Used by permission of Penguin Books, an imprint of Penguin Publishing Group, a division of Penguin Random House LLC. All rights reserved.

Excerpt from *To Speak for the Trees: My Life's Journey from Ancient Celtic Wisdom to a Healing Vision of the Forest* copyright © 2019 by Diana Beresford-Kroeger. Published by Timber Press in paperback and originally in hardcover by Random House Canada. All rights reserved. By permission of Stuart Bernstein Representation for Artists, New York, NY and protected by the

How to Do Nothing: Resisting the Attention Economy by Jenny Odell quoted here courtesy of Melville House Publishing, LLC.

Limits to Growth: The 30-Year Update by Donella Meadows, Jorgen Randers, and Dennis Meadows quoted here courtesy of Chelsea Green Publishing Company.

INDEX

shift in, 72
shifting in rituals, 168
Conversation, the, 8–9, 14–15, 205
Cook, Francis H., 50
courage, 7, 106, 146, 188
anger and, 142–43
of bodhisattvas, 10
compassion and, 123
maintaining, 33
need for, 117–20
COVID-19 pandemic, lessons of, 169–70
craving, 54, 60–64, 73, 102
culture of enough, 193
cynicism, 177–78, 179, 202

Dalai Lama. *See* Tenzin Gyatso, the
    Fourteenth Dalai Lama
death
    of author's mother, 175–76
    denying, 171
    reflecting on, 155
decarbonization, 78, 100, 213
deforestation
    in Amazon, 40–41
    in Kenya, 82, 83
    pollinators and, 87
    in Tibet, 7
delusion, 57–58, 60, 73, 102
Descartes, René, 52, 53
despair, 4–5, 73, 124, 125, 150–51, 220
Dharamsala, 1
Diné (Navajo) tradition, 90
disliking, 54, 64–66, 73, 102
Don't Look Up (movie), 135, 207, 224–25
drought, 138, 185
    forests and, 41
    global warming and, 40
    jet stream and, 35
    as opportunity, 106–7
    pollinators and, 87
Duffy, Philip, 79
Duvanny Yar (Russia), 30

Earth
    and body, relationship of, 53
    cooling, 45

desacralization of, 53
healing capacity of, 13, 79–80, 100
honoring, 202–3
humans and, 60, 94
taking for granted, 171
See also global warming
Earth Emergency film, 163–64
East River, 28
ecoanxiety. *See* climate anxiety
ecodestructive society, 139
Ecological Revolution, 72
economy
    changing, 71–73
    global, 46
    political, 61–62
    regenerative, 96
ecopsychology, 139
education and knowledge, 13, 22, 60,
    205–8
Eisenberg, Nancy, 142
Emanuel, Kerry, 24, 25, 37
emissions, 55
    cutting, limitation of, 24
    fossil-fuel, 25, 39, 104, 213–14
    industry-driven, 187
    population growth and, 213
    reporting requirements, 58
    zero goal, 115–16
    See also greenhouse gases
emotions, 68–69, 150–51
endangered species, 103, 183–84
energy consumption, 66
Environmental Research Letters, 212–13,
    261n27
envisioning, 177–79
Epel, Elissa, 38, 55–56, 220–22
equanimity, 141, 144, 200
Ernman, Malena, 195
ethics, 13, 100–101, 141, 204, 207. *See also*
    moral suffering, four types
Evers, Medgar, 142
existential crisis, 32, 135–36, 137, 139, 155–56,
    171–72

faith communities, 191–93, 198
farming, 57, 85, 100, 201

fear, 64–65, 117–18, 145, 146
  of climate change, 138, 206–7
  compassion and, 122
  in meditation, working with, 69–70
feedback loops, 72, 230. *See also* climate
      feedback loops; human
      feedback loops
Figueres, Christiana, 126–27, 208–10
Fisher, Andy, 139
Foer, Jonathan Safran, *We Are the Weather*,
      204
food forests, 91–92
forests, 84
  boreal, 41–42
  functioning of, 39–40
  importance of, 44–45
  kelp, 90–91
  old-growth, 77, 92, 101
  old-growth vs. new-growth, 43–44
  protecting ("pro-forestation"), 45
  temperate, 42–43
  tropical, 40–41, 84–86
  See also deforestation; food forests
fossil-fuel industry, 121, 178, 187, 229
fractals, metaphor of, 188
Francis, Jennifer, 33–35, 36–37, 125
Francis, Saint, 107

Garrigus, Beth, 144
generosity, 113, 120, 188
Germany, 143, 159
Gill, Charlotte, 43
glacial melting, 27–28, 125, 230
global bioplan, 88–89
Global Carbon Atlas, 55
Global Development and Environment
      Institute, 44
Global Environments Network, 147
global warming, 23, 26–28, 30, 39, 43
global weirding, 35
Goodall, Jane, 222–23
grandmother's heart, 144
Great Plains, 128–29
Great Turning, 71–73, 107
greed, 16, 61, 151
Green Belt Movement, 83–84

greenhouse gases, 29, 30, 140
  in climate feedback loops, 23
  effects of, 26, 33–35, 39, 80
  reducing, 35, 97–98, 229
  top ten polluters, 59
Greenland, 27–28
grief, 14
  as bridge, 150–51
  fear of, 147–48
  in meditation, working with, 69–70
  ritual honoring, 144–46
  stages of, 139, 140, 156
  toolkit for, 147–49
Guenther, Genevieve, 213, 219–20

Halifax, Roshi Joan, 71, 140–42, 144, 185–86
Hall, Carl, 163–64
happiness
  materialism and, 117
  of others, 196–97
Harari, Yuval Noah, 16, 225
Hartmann, Thom, 154
Hawken, Paul, 78–79, 212
  Blessed Unrest, 86
  Regeneration, 79, 206, 214
  regeneration guidelines, 213–14
Hayhoe, Katharine, 35, 120–22, 128, 210
health, stable climate and, 108
heartbreak, 135, 156
  habits and, 134
  metabolizing, 147–51
  ritual for, 144–46
  sharing, 136–39
Heglar, Mary Annaïse, 225
Heiltsuk Nation, 90–91
helplessness, 4–5, 56, 60, 138, 143
Hershey, Barry, 5, 9
Higgs, Stephanie, 78–79, 147
Hockney, David, 166
Holland, Marika, 27
honesty, 12, 46, 119, 139, 178, 226
hope, 127–28
  absolute and active, 221–22
  capacity for, 123–26
  maintaining, 33
  need for, 105

hopelessness, 58, 105
hospice metaphor, 71
human feedback loops, 54–55, 104, 105,
    206–7
    communication and, 211
    craving, 60–64, 112–13
    delusion, 57–58, 60
    disliking and fearing, 64–66
humans, 54
    capacity to change, 104–7, 108, 122–23
    community of, 51
    dominance of, 154, 195
    extinction of, 155–56
    impact of, 181
    interconnectedness of, 196
    as keystone species, 129–30
    as part of nature, 78–79
    responsibility of, 33, 59, 75, 111, 143
    shared humanity of, 118, 197
    systemic collective agreements of, 151
    as troublemakers, 19, 47, 118
hurricanes, 28, 37

iceberg model, 189–90
Iceland, 230
ignorance, 46, 54, 57–58, 109, 206
illness, 137, 171–72
impermanence, 150–51
India
    clean water in, 7, 109–10
    environmental movements in, 52, 122
    Tibetan exile community in, 114–15,
    116
Indigenous peoples, 52, 90–93, 129–30, 199
Indra's Net, 50, 94, 154
Industrial Growth Society, 61–62, 101, 171
    alternative to, 62–64
    being productive in, 162–63
    exploitation of, 142
    fear and, 64–65
    grief and, 148
    nature and, 199, 201
    rituals of, 149
    transitioning from, 71–73
Inflation Reduction Act, 229, 230
Ingram, Catherine, 137, 154, 155, 217–18

interdependence, 11, 130, 181
    change and, 108–10
    craving and, 60, 61
    denying, 52, 54
    in Diné language, 106
    growth and, 63–64
    implications of, 187–91
    importance of understanding, 14
    with nature, 198–202
    neglecting, 196
    recognizing, 51
    science and, 79–80, 192
    See also Indra's Net
Intergovernmental Panel on Climate
    Change (IPCC), 9, 21, 57, 111,
    122
Islam, 120

Jataka tale, 190
jet stream, 24–25, 33–36, 80
Jinpa, Thupten, 12, 161–62
    on bodhisattvas, 119–20
    on compassion and courage, 123
    on fear, 117–18
    Fearless Heart, 119
    on interdependence, 50–51
Johnson, Ayana Elizabeth, 215–16
joy, 101, 147, 172, 173, 184, 196, 197, 222
Judaism, 120
June, Lyla, 105–7, 199
    on eating, connection in, 200
    "Go to the Earth in the Morning,"
    202–3
    on hope, 128–30
    on Indigenous models of
    sustainability, 90–93

Kangchenjunga (mountain), 198
Kangyur Rinpoche, 160
Kanko, Kritee, 144–46, 148
karma, 13, 46, 51
Karuna-Shechen organization, 160
Karura Forest, 84
Keltner, Dacher, 159
Kenya, 81–84, 161, 198
King, Martin Luther, Jr., 142

Solnit, Rebecca, 123–24, 126, 170–71, 178, 191, 225
stewardship, 89, 90–93, 129–30
sub-Saharan Africa, 122
suffering, 4, 57
    carrying, 67
    of others, relieving, 10
    repressing, forced, 139
    root of, 54
    from self-centeredness, 196–97
    speaking about, 136–39
    with world, 136, 140
Sunrise Movement, 186–87, 219
super-blocks, 98–99
sustainability, 72
    of cities, 95–99
    of consumerism, standards for, 61
    in fishing, 91
    interdependence and, 187
    possibility of, 45–47
    vision of, 65
Sweden, 7–9
systemic change, personal transformation and, 189
systems theory, 46, 50, 61, 179. See also iceberg model

Tade, Stephanie, 133–35, 178
technology, 225
    limitations of, 63, 100–102
    mobility and, 170–71
Tenzin Gyatso, the Fourteenth Dalai Lama
    on anger, 142–43
    author's experience with, 49
    biographical information, 6–7, 8
    on change, 10–11, 181
    on community, 131
    on Earth, 199
    and Greta Thunberg, first meeting, 1–3
    on human beings, 19, 47
    on human responsibility, 75
    on humans and environment, link between, 202
    on materialism, 114–15
    nature programs, fondness for, 161–62

"One Humanity Practice," 196–97
    on optimism, 31–33, 126
    on reforestation, 80
    on technology, 101–2
    three poisons, 55, 73. See also craving; delusion; disliking
Thunberg, Greta, 1, 184, 195
    activism of, 8, 21
    Atlantic sailboat crossing, 162, 258n7
    on awareness, 135
    background of, 8–9, 54
    on beginning of the beginning, 227
    on buying and building, 100
    on climate justice, 115–16, 121
    on communication, 19
    on community, 131
    and Dalai Lama, first meeting, 1–3
    on failure, 130
    on fairy tales of growth, 62
    family to-do list, 223–24
    on hope, 123
    on living locally, 170–71
    name, pronunciation, 21
    on nature, rewilding, 75
    popularity of, 225–26
    school strike of, 119
    on social movements, creating, 181, 226
    "Speak the Truth," 110–12
    Trump and, 140
    on waking up, 104
    See also Our House Is on Fire (Thunberg and Ernman family)
Thunberg, Svante, 193–94
Tibet, 6–9
Tibetan Plateau, 7, 8, 30, 192
tipping points, 22–23, 24, 41, 42
to-do lists, 185
    A. E. Johnson and K. K. Wilkinson's, 215–16
    Catherine Ingram's, 217–18
    Donella Meadows's, 218–19
    Elissa Epel's, 220–22
    Five Things, 212–13, 261–62n27
    Genevieve Guenther's, 219–20
    Jane Goodall's, 222–23

Paul Hawken's, 213–14
Thunrberg-Ernman Family's, 223–24
tonglen practice (giving and receiving), 68
transportation
air travel, 104, 166, 184, 191, 193–94
imbalance of, 162
mass transit, 95
trauma, ecological, 200–201
travel and mobility, 166, 170–71. *See also*
wildlife tourism
trees
healing power of, 87
necessity of, 38–39
planting, 88–89, 101
and streams, interdependence of,
81–83
See also forests
Tsétsêhéstâhese (Cheyenne) tradition, 90
Tucson, 87, 98–100

uncertainty, 49, 117, 124, 221
United Nations
Climate Action Summit, 110–12
Conferences on Climate Change, 122
Framework Convention on Climate
Change, 126
Thunberg's address to, 21, 24, 62
United States, 30
United States Congress, 21

values, 102, 188–89, 193, 199, 210
virtue signaling, 191

Wallace-Wells, David, 213
Waltch, Bonnie, 124–25
Wamsler, Christine, 187, 189–90
Washington, Warren, 36, 37
water restoration, 192
water vapor feedback, 36–37, 40–41
weather-associated disasters, 3–4, 28, 35,
106
Westervelt, Amy, 225

wetlands, 77–78
Weyler, Rex, 200–201
Whyte, Kyle, 93–94, 198
wild rice, 93–94
wildfires
in boreal zone, 41–42
controlled burning and, 128–29
global warming and, 40
jet stream and, 35
Wildlife Alliance, 85
wildlife tourism, 85, 164, 166
Wilkinson, Katherine K., 215–16
witches, 52–53
wonderment
of climate scientists, 165
as envisioning, 177
examples of, 172–74
at home, 170
importance of, 158–59
Industrial Growth Society and,
163
inspiration from, 162
types and effects of, 159–61
where we are, 166–68
wood pellet industry, 42–43, 44
Woodwell, George, 23, 39, 40, 81
Woodwell Climate Research Center, 28,
29–31, 33, 40, 79
Woolf, Virginia, 123–24
World Resources Institute, 44
Wu, Michelle, 98

Yale Program on Climate Change
Communication (YPCCC),
56–57
youth and children
activism of, 2–3, 16
as natural visionaries, 177–78
outlook of, 56

Zen, 144
Zuniga, Adriana, 87

ABOUT THE AUTHORS

SUSAN BAUER-WU is the former President of the Mind & Life Institute, an organization cofounded by the Dalai Lama in 1987 to bring science and contemplative wisdom together to better understand the mind and create positive change in the world. In her work with Mind & Life, she has championed "human-earth connection" as a priority. She began her career as a registered nurse specializing in oncology and end-of-life care, and later completed PhD studies in psychoneuroimmunology. She has held leadership, teaching, and clinical positions in nonprofits, higher education, and health care, and is the author of *Leaves Falling Gently: Living Fully with Serious & Life-Limiting Illness through Mindfulness, Compassion & Connectedness*. In her free time, she is outdoors as much as possible, gardening and hiking the Blue Ridge mountains.

STEPHANIE HIGGS writes, edits, and publishes in the East Village, Manhattan, where she lives with her husband and dog. She is the cofounder of Two Shrews Press and coauthor of *Beyond Addiction: How Science and Kindness Help People Change*.